T0321949

NEWT🍎N'S CHICKEN

Science in the Kitchen

World Scientific Series on Science Communication

Editor-in-Chief
Hans Peter Peters
Forschungszentrum Jülich, Germany
h.p.peters@fz-juelich.de

Aims and Scope

The books published in this series deal with the public communication of science, i.e. with communication of and about science involving non-scientists and taking place in the public sphere. Possible topics include public communication and discourses about scientific knowledge, scientific projects or research fields, and communication about science as a social system and its interdependencies with the larger society.

This book series is open to analyses of all forms of public communication and interaction: journalism, public relations of science, blogs, social media, video-sharing platforms, science museums, public events, engagement activities, public deliberation and participation, citizen science and other collaborations between scientists and citizens, for example. Books may focus on content and processes of messages and discourses, on actors and their strategies, on channel and arena characteristics, on the reception, effects and use of public expertise, on public controversies, on inclusion of citizens in public discourses, and on issues of quality, ethics, and trust.

Typically, authors/editors will come from the academic field and have an academic audience in mind. But some of the books may also be relevant for communication professionals, scientists (as communicators), science managers and knowledge users, for example. Books may be based on specific research projects, deal with a relevant subjects by means of review of existing studies and theoretical discussion, or publish contributions of a relevant conference (proceedings).

Published

NEWTN'S CHICKEN

Science in the Kitchen

A menu which stimulates the taste buds, an appetite for
scientific knowledge and the love of reading.

Massimiano Bucchi

University of Trento, Italy

World Scientific

NEW JERSEY · LONDON · SINGAPORE · BEIJING · SHANGHAI · HONG KONG · TAIPEI · CHENNAI · TOKYO

Published by

World Scientific Publishing Co. Pte. Ltd.

5 Toh Tuck Link, Singapore 596224

USA office: 27 Warren Street, Suite 401-402, Hackensack, NJ 07601

UK office: 57 Shelton Street, Covent Garden, London WC2H 9HE

Library of Congress Cataloging-in-Publication Data
Names: Bucchi, Massimiano, 1970–   author.
Title: Newton's chicken : science in the kitchen / Massimiano Bucchi, University of Trento, Italy.
Other titles: Il pollo di Newton. English
Description: 1st. | New Jersey : World Scientific, [2021] | Series: World scientific series on
    science communication ; Volume 2 | "Originally published in Italian as Il Pollo di Newton:
    La scienza in cucina, 2013 Ugo Guanda S.p.A., Parma"--Title page verso. |
    Includes bibliographical references and index.
Identifiers: LCCN 2020041562 (print) | LCCN 2020041563 (ebook) |
    ISBN 9789811225444 (hardcover) | ISBN 9789811225451 (ebook)
Subjects: LCSH: Science--History. | Cooking--History. | Science--Miscellanea. |
    Cooking--Miscellanea.
Classification: LCC Q125 .B9218 2021 (print) | LCC Q125 (ebook) | DDC 641.501/5--dc23
LC record available at https://lccn.loc.gov/2020041562
LC ebook record available at https://lccn.loc.gov/2020041563

British Library Cataloguing-in-Publication Data
A catalogue record for this book is available from the British Library.

Originally published in Italian as *Il Pollo di Newton: La scienza in cucina,*
© 2013 Massimiano Bucchi. English text translated by Tania Aragona.

Copyright © 2021 by World Scientific Publishing Co. Pte. Ltd.

All rights reserved. This book, or parts thereof, may not be reproduced in any form or by any means, electronic or mechanical, including photocopying, recording or any information storage and retrieval system now known or to be invented, without written permission from the publisher.

For photocopying of material in this volume, please pay a copying fee through the Copyright Clearance Center, Inc., 222 Rosewood Drive, Danvers, MA 01923, USA. In this case permission to photocopy is not required from the publisher.

For any available supplementary material, please visit
https://www.worldscientific.com/worldscibooks/10.1142/11970#t=suppl

Printed in Singapore

MENU

STARTER, FOLLOWED BY FIRST COURSE

Cookery as Science, and Science as Cookery: from Socrates to Cold Fusion

(Bacon) asked them what they would take for their draught; they answered so much: his Lordship would offer them no more but so much. They drew-up their nets, and in it were only 2 or 3 little fishes: his Lordship then told them it had been better for them to have taken his offer. They replied, they hoped to have had a better draught; "but" said his Lordship, "Hope is a good breakfast but an ill supper."

JOHN AUBREY
Brief Lives

Starter: Mayonnaise, Science and Women

Science in the Kitchen is the title of a section that has been featured for years in the longest-living and most viewed Italian television science programme: *SuperQuark*, conducted by Piero Angela.[1]

The section invites viewers to discover the physical and chemical processes that take place within the kitchen. The first episode is dedicated to mayonnaise. The section starts with a series of interviews of people on the streets, at the market or in the gym, in which their knowledge of mayonnaise is put to test: do they know which ingredients are required and how to make mayonnaise? The questions are short and often out of context (for example "how often do you make it?" the interviewer asks a young girl doing a workout at the gym), but many answers reveal people's ignorance or complete confusion. A bus driver for example mentions "fermentation"; a person at the market talks about "parsley" and "flour". Others admit to not remembering any of the ingredients. Some questions mention sayings such as the influence of weather conditions, moon phases or even women's menstrual cycles on the success of a good mayonnaise.

When he returns to the television studio, Piero Angela comments: "well, it seems everyone has his or her own theory, and we have ours". Before embarking on his explanation for "the science of mayonnaise" however, Angela shows the viewers a candid camera clip. The clip shows a group of women attending a cooking class at the home of an expert cook. The women are unsuccessfully trying to make mayonnaise, but the mayonnaise isn't having any of it and keeps "curdling". In the end, the candid camera is revealed much to the relief of the aspiring cooks.

---

[1] Broadcasted since 1981 in different formats in prime time, the programme routinely ranks among most viewed programmes in Italy, with audience peaks over 5 millions.

"We have included a small trick to force the mayonnaise to turn", the presenter continues, "but first let's see what happens when mayonnaise is made normally". With the support of an animated film, the viewers learn that mayonnaise thickens thanks to the presence of lecithin, a substance contained in egg yolk that enables the water (contained in the lemon and the eggs) and the oil to mix. Lecithin surrounds the oil droplets and thus stops them from mixing together, bonding them to the water. In the case of the cookery school, the women were tricked by the addition of an anti-emulsion substance.

The *Science in the Kitchen* section is not just unusual; it is also profoundly revealing of a strategy — an ideology even — for presenting science in public.

Science has acquired a preeminent role in contemporary society and as such it is constantly engaged in legitimising and reinforcing its relevance. Of the strategies for legitimising science in the public arena, two are the most common. The first is linked to its utility: science justifies its role through the presentation of the benefits of its applications and consequences particularly in the field of technology. The second stresses its cultural importance: science thus becomes a source of cultural enrichment, aesthetic pleasure or even entertainment. This is a tradition that dates back to the highly successful public conferences at the Royal Institute during the early XIXth century and to the great fairs and exhibitions at which the most recent developments in the realm of science and technology left visitors wide-eyed. The same happens today when we are presented with spectacular images from astronomical or particle physics observations.

Science in the Kitchen can be seen as a variation of this strategy based on the entrance of science and its methods into daily life. Instead of offering wonderful or extraordinary content, science works its way into everyday experience, explaining the mechanisms that govern queues at the supermarkets, the physical and mathematical secrets of football, or the reasons for — as the case may be — mayonnaise curdling or emulsifying.

Within this strategy, scientific knowledge is not presented in opposition to common sense nor does it attempt to subvert it as has become typical for some forms of science popularisation, especially since the huge public impact of XXth century revolutions in physics. "To go beyond the experience of everyday life" was one of the tasks for science according to physicist Hermann Bondi (1964, p. 62); for the philosopher Gaston Bachelard, the development of science was marked by a progressive discontinuity and distancing from common sense and daily experience (1938).

Here, on the contrary, science sits side by side with common sense, ready to take it by the hand and elevate it by enlightening the meaning of consolidated practices. After all, as the conductor Piero Angela comments at the beginning of the programme, "cooks are, in their own way, the inventors of chemical reactions".

Here we have a more subtle rhetorical strategy. Common sense is not contradicted but rather ridiculed like an indigenous heap of unwitting rituals that verge on magic, for which only an external, scientific observer can provide a satisfying explanation. The people interviewed on the street are an almost literal representation of common sense ("the man on the street"), to whom questions are asked that are often deliberately incomplete and misleading in order to underline their ingenuity. And how entertaining, naïve and ultimately pathetic do the students from

Fig. 1.1 Kitchen Laboratory, www.clementoni.it.

the cooking class appear to us, tricked so easily by a simple chemical subterfuge!

This way science extends its authority and analytical methods to parts of life and its practicalities such as cooking and housekeeping that were traditionally governed by common sense, accompanying the man on the street towards this newly acquired awareness with a touch of paternalism, as if he were an innocent finally emerging out of superstition.

Today, using the kitchen and its secrets to present and popularise science has become common practice. Popular books, installations in science centres, television and radio programmes and children's games invite us to discover these secrets, offer "recipes for having fun with science", "experimental-recipes to learn about science and nutrition" and even "epicurean laboratories in which to explore the science of food preparation".[2] But the presentation of cooking as a science is nothing new.

---

[2] Just a few examples: Susan Strand Noad, *Recipes for Science Fun*, Watts, New York, 1979. Julia B. Waxter, *The Science Cookbook: Experiment recipes that teach science and nutrition*, Fearon, Belmont, 1981; Tina L. Seeling, *The Epicurean Laboratory: Exploring the science of cooking*, Freeman/Scientific American, New York, 1991; id., *Incredible Edible Science*, Freeman/Scientific American, New York, 1994; Peter Barham, *The Science of Cooking*, Springer, Berlin, 2001 Robert L. Wolke, *What Einstein Told His Cook*, Norton, New York, 2002, *La matematica in cucina* (*Mathematics in the Kitchen*), Bollati Boringhieri, Torino, 2004. Many texts offer activities for children, such as K. Woodward, *Science in the Kitchen*, EDC, 1992. One work that is more focused on scientific analysis and has become a classic reference is Harold McGee, *On Food and Cooking, The Science and Lore of the Kitchen*, Scribner, New York, 1984, Italian transl. *Il cibo e la cucina. Scienza e cultura deglia alimenti*, Franco Muzzio, Padova 1989. The popular BBC weekly radio program *The Naked Scientists*, conducted by a group of physicists from Cambridge provides a wide section on the Internet dedicated to scientific experiments that take place in the kitchen (http://www.thenakedscientists.com/html/content/kitchenscience/). In Italy, the column *Pentole e Provette* (*Pots and Test Tubes*) by chemist Dario Bressanini is published on the Italian edition of the magazine *Scientific American* and on the author's blog; the website of the radio program on science *Moebius*, conducted by Federico Pedrocchi on Radio 24 includes a blog on scientific cooking which the participation of physicist Davide Cassi, an expert in molecular cooking (cf. chap. *Dessert*).

Cookery as Science,
Or how Liebig stirred up housewives against doctors and chemists

Since the XVIIth century, food preparation and household chores have been presented as sophisticated technical activities worthy of "scientific" knowledge. In the ensuing years, frequent references were made to the "surgery" and dissection of chickens and pigs and advice was given to help the domestic goddess perfect her competence in the sphere of "physics in cookery" and medicine (cf. for example Markam, 1615; Cock, 1675).

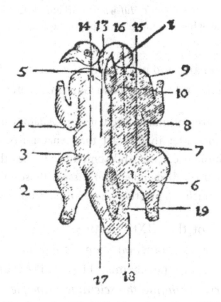

Fig. 1.2 *The dissection of a boiled hen*, from Rose (1682, p. 39).

Fig. 1.3 Frontispiece of *The English Huf-wife* (Markam, 1615), in which the author stresses the importance of being competent in "physics and cookery".

In his treatise on arts and professions, *La Piazza universale di tutte le professioni del mondo* (1585), Tommaso Garzoni wrote:

> Butchers (…) are not much different from Anatomists, and only differ from the latter in the fact that Anatomists dissect and dismember human corpses and sometimes bodies that are still alive, but butchers dismember and dissect those of beasts and animals with much less pity than those in an Anatomy laboratory.[3]

During the course of the XIXth century, the focus on cooking as a science became a real phenomenon from a scientific, cultural and social point of view. In 1821, the chemist Friedrich Accum published his *Culinary Chemistry, exhibiting the scientific principles of Cookery, with concise instructions for preparing good and wholesome Pickles, Vinegar,*

---

[3] Garzoni (1585, ed. 1588, p. 152). On the sharing of tools, techniques and places between food preparation and early experimental philosophy, see also Guerrini (2016).

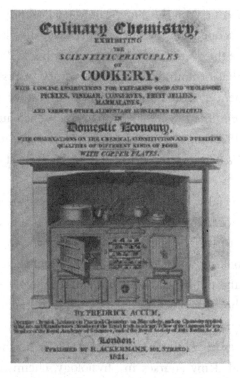

Fig. 1.4 Title page of F. Accum (1821), *Culinary Chemistry*.

Conserves, Fruit Jellies with observations on the chemical constitution and nutritive qualities of different kinds of food.

In his *Essays, moral, philosophical and stomachial on the important science of good living* (1822), a classic that is reprinted to this day, Lancelot Sturgeon (pseudonym) claims that "a cook, to be thoroughly accomplished, should not only be deeply versed in all the arcane of the kitchen, but should possess an intimate knowledge of ichthyology, zoology, anatomy, botany and chemistry" (p. 190).

The influence of German chemist and entrepreneur Justus von Liebig (1803–1873) is of great importance for numerous books on cooking and domestic economy published from the mid XIXth century such as the *Manual of Domestic Economy* by John Henry Walsh (1856) or the *Manual of Cooking and Chemical Economics* by Hermann Klencke (1857). According to the latter, "cooking is nothing more than a chemistry

9

laboratory"; that same year, the American Mary Peabody Mann hopes to soon see a microscope in every kitchen, because, she claims, "chemical analysis should be the guide for of every cookery book".[4]

Two books that profoundly mark the history of taste and cooking also refer to science, although they do so in rather different ways: *Physiologie du goût* (*The Physiology of Taste*) by French author Anthelme Brillat-Savarin (1825) and *La scienza in cucina e l'arte di mangiar bene* (*Science in the Kitchen and the Art of Eating Well*) by Italian Pellegrino Artusi (1891).

But it is especially during the last decades of the XIXth century and the early decades of the XXth century that a flourish of publications promise to bring "science into the kitchen" and to reveal the "chemistry of cooking and cleaning".[5] Well-established scientists hold cooking classes and published detailed manuals, mostly aimed at female audiences, on the properties of foods and the reactions that occur during their preparation. Albert J. Bellows, a chemistry and physiology professor dedicates his *Philosophy of Eating* (1867) "to the five thousand ladies who attended my courses in physiology, chemistry and hygiene between 1838 and 1858".

---

[4] Cit. in in Finlay (1995, p. 55). One should at least remember the series of *Familiar Letters on Chemistry* by Liebig published in various editions and formats and translated in numerous languages from 1851. Highly significant within this scenario is also the impact of new inventions and techniques for the preparation and the conservation of foods, that intensified from the end of the 17th century and especially the following ones. For a broader overview see the chapter on "Science and Technology in the kitchen" in Capatti and Montanari (1999, pp. 285–319).

[5] Other notable examples in English include: *The Science of Nutrition* (Atkinson, 1892); *The Chemistry of Cooking and Cleaning* (Richards, 1882); *The Chemistry of Cookery* (Williams, 1885); *Science in the Daily Meal* (Broadbent, 1900); *Principles of Cooking: A textbook in domestic science* (Conley, 1914). Among the many magazines: "Mother's Friend", "Hearth and Home", "The Ladies' Home Journal" and "The Boston School Magazine of Culinary Science and Domestic Economics". In Italy among others: Paolo Mantegazza, *Enciclopedia igienica popolare. Igiene della cucina (1866)* (*The Popular Encyclopaedia of Hygiene: Hygiene in the kitchen*); Oscar Giacchi, *Il medico in cucina, ovvero perche si mangia e come dobbiamo mangiare (1882)* (*A Doctor in the Kitchen or Why We Eat and How We Should Eat*); Amedeo Pettini, *Dall'empirismo alla cucina scientifica (1905)* (*From Empiricism to Scientific Cooking*).

In the preface to his *Handbook of Practical Cookery for Ladies and Professional Cooks. Containing the Whole Science and Art of Preparing Human Food* (1867), Pierre Blot writes: "A cook-book is like a book on chemistry, it cannot be used to any advantage if theory is not blended with practice" (Blot, 1867, p. 3).

Fig. 1.5 Title cover of Ervilla Kellogg's *Science in the Kitchen* (1892). One should note the still and the microscope in the top right corner.

Ella Ervilla Kellogg introduces her *Science in The Kitchen. A Scientific Treatise on Food Substances and their Dietetic Properties together with a Practical Explanation of the Principles of Healthful Cookery and A Thousand Choice, Palatable and Wholesome Recipes* (1892) as follows:

It is a singular and lamentable fact, the evil consequences of which are widespread, that the preparation of food, although involving both chemical and physical processes, has been less advanced by the results of modern researches and discoveries in chemistry and physics than any other department of human industry (...) the art of cookery is at least a century behind in the march of scientific progress. The mistress of the kitchen is still groping her way amid the uncertainties of medieval methods, and daily bemoaning the sad results of the "rule of thumb". The chemistry of cookery is as little known to the average housewife as were the results of modern chemistry to the old alchemists: and the attempt to make wholesome, palatable and nourishing food by the methods commonly employed, *is rarely more successful than that of those misguided alchemists in transmuting lead and copper into silver and gold.* The new cookery brings order from the confusion of mixtures and messes, often incongruence and incompatible, which surrounds the average cook, *by the elucidation of the principles which govern the operations of the kitchen, with the same certainty with which the law of gravity rules the planets.* (Kellogg, 1892, p. 3, my italics)

The plan in these texts is therefore very clear: to bring the scientific revolution into the kitchen, by replacing outdated and incongruous practices just as science had done with alchemy; and to achieve the precision and regularity of processes that science has discovered in physics and astronomy to shine likewise in the kitchen.

One of the most famous chefs of those times, the head chef of the Savoia Family Amedeo Pettini, summarises it well in his book *Dall'empirismo alla cucina scientifica From Empiricism to Scientific Cooking* (1905).

Do observe — kind Sires — how everything nowadays is being organised and harmonised according to scientific criteria, and tell me if cookery could possibly be the only area to escape this trend! (...) Scientific cookery, that will become the conquest of our century, will eliminate these aberrations. The cook, (...) recognised solely for his capacity to prepare... indigestions, will no longer hold this role. He will take pride, having become an expert in scientific knowledge, in setting a truly physiological table, not inferior but rather superior — in all aspects — to the empirical one of our days. (Pettini 1905, pp. 150 and 156)

Magazines such as "Mother's Friend", "Hearth and Home", "The Ladies' Home Journal" and "The Boston School Magazine of Culinary Science and Domestic Economics" offer an impressive quantity of technical information, often extremely detailed, on the properties of foods and the most appropriate ways of preparing them.

Mrs Kate Hunnibee, the author of a popular advice column for housewives in *Hearth and Home*, informs her readers of her indecision between cooking a roast beef or a roast chicken. In the end, her hesitation was overcome based on the advice from "some famous experts who considered chicken to be more nutritious compared to roast beef, whilst others claimed roast beef to be more substantial; but chicken could be digested an hour sooner than roast beef". The rest of the menu was also based upon considerations of a technical and scientific nature. "Blueberry sauce contains no nutrients other than sugar, but it helps absorb everything else (...), potatoes, pasta, bread, butter and tapioca can provide the necessary starch".[6]

This vast quantity of literature and publications stun the enthusiastic adepts of science *Bouvard and Pecuchet* in Flaubert's novel (1881), when they decide to base their respective diets on the advice of the most reputable doctors and scientists of their time.

> Every kind of meat had its inconveniences. Puddings and sausages, red herrings, lobsters, and game are "refractory". The bigger a fish is, the more gelatine it contains, and consequently the heavier it is. Vegetables cause acidity, macaroni makes people dream; cheeses, "considered generally, are difficult of digestion". (Flaubert, 1881, Eng. trans. 2013, p. 74)

By now fully recognised both at the institutional and public level, science becomes a resource and an element of legitimation for other practices. To be "scientific" — to describe this movement in a rather simplistic way — is very fashionable between the XIXth and XXth century and it is considered *chic* to refer to science, even in gastronomy.

---

[6] Mrs Kate Hunnibee's Diary in "*Hearth and Home*", I, 1869, p. 108.

Some scholars have underlined how this flourishing of the "science of cooking" and "domestic economy" was also responding to the need of offering a scientific context — albeit a defined and bounded one — to the increasing number of women who were eager to professionally access science and medicine.[7] So "domestic" sciences became an area of "domestication" for the emerging demand for engagement within science by female audiences.

Nevertheless, there were several occasions in which these audiences played a very active role with regard to scientific debates. Support from housewives, for example, was essential to promote the market introduction and consolidation of meat extracts by German chemist and entrepreneur Justus von Liebig. From 1865, numerous scientists and doctors had expressed their scepticism for the new product, casting doubts on its nutrition properties. According to physician Edward Smith, meat extract was so poor compared to meat that it resembled "a play by Hamlet without the character of Hamlet"; according to physiologist Ernst Kemmerich, the potassium salts contained in the extract were even potentially toxic. Yet, Liebig was able to redefine the public interested in his products — which had originally been aimed mainly at hospitals and chemists — and could now target domestic kitchens; the emphasis of his communication insisted more on the practical and economic aspects of his product and less on the nutritional values of the extracts. "Incapable of persuading doctors and chemists, he tried to persuade the consumers" (Finlay, 1992, p. 61). Liebig was keen to cultivate his public of housewives, to whom he regularly offered cooking manuals, advertising pamphlets and his famous "trade cards" (signed on the back by the scientist, as a quality control guarantee of the products) that soon became passionately collected items. By offering professional contracts in order to reach his public, Liebig sought the support of the most influential female authors of books on cookery and news stories on "domestic science". This audience quickly and enthusiastically adopted Liebig's innovative invention. They saw meat extract as an efficient resource to

---

[7] See the Apple example (1995).

Fig. 1.6 Liebig postcard, series 737, 1903.

manage their cooking activities; they also considered it an opportunity to selectively incorporate rationality and scientific authority in forms congruent with their daily expectations and habits.

Science as Cookery,

or how a revolution in geology risked being mistaken for a minestrone

Having described how cookery has been connected to and compared to science, we can now explore the reverse: how the practice of science has been touched by cookery.

"Don't worry, you just have to follow the recipe. You'll see, it's just like cooking": this is how the molecular biologists of the Max Planck institute in Monaco used to encourage me and other guests who were not experts during laboratory training (incidentally, they were also in the habit of preparing coffee using the same hot-plate they used to heat their test tubes).

For the sociologist of science Trevor Pinch, the best way to appreciate the importance of the tacit skills in science particularly in laboratory practices is to try... cooking a cheese soufflé! According to Pinch, this is an almost impossible task if one tries to make it solely on the basis of the written recipe without having had the chance to witness "first hand" the practical details that make all the difference (gestures, choice of materials, minor adjustments). When we try to cook a new dish based on scant instructions provided by a famous chef on television, we find ourselves in a very similar situation to those researchers and technicians who operate in areas at the edge, trying for example to replicate new or uncertain results. If the result is unsatisfactory we ask ourselves: where did I go wrong? Did I miss a crucial step? Or could it be that the famous chef forgot a significant detail? Or perhaps this dish is not as good as it appeared to be? These are similar questions to the ones asked by those who try unsuccessfully to replicate an experiment or scientific result that has yet to be consolidated: did I make a technical mistake? Or is the description by my colleagues in their paper incomplete? Or was their result ultimately incorrect? At the end of the 1960s, for example, many laboratories tried to develop new models of gas lasers with increased power. After a series of attempts, a research laboratory linked to the Canadian defense sector developed an innovative model called the TEA Laser (for Transversely Excited Atmospheric Pressure Laser), which was very different to the one originally designed, and the personnel took a few months to study the actual basics of how it worked. The Canadian group published its own results in an article in a specialised journal and presented its results at some conferences. But for almost two years, efforts by other laboratories to build an identical laser based on the articles and presentations lead to nothing. Building other similar lasers only became possible after a long series of meetings between research groups, visits by researchers and technicians to third party laboratories, and many exchanges of materials and instruments. In other words, it became possible only once the vast amount of tacit knowledge that is difficult to verbalise and is often crucial even in the most complex technological sectors was shared.

From a purely formal point of view, it is easy to see a certain similarity between the most sophisticated culinary recipes and some experimental protocols in contemporary biology: in both cases, there is a description of the "ingredients" (materials and instruments), followed by instructions, at times combined with precise indications of quantities ("blend 200 g for 3 minutes"), which often count on the experience and the manual expertise of the experimenter to "roughly" calculate the quantities.

> *Quickly* melt the vial (...) and add the *appropriate quantity* of culture (depending on the number of cells present in the vial) (...) inspect the cultures every day to determine the *optimal density* for the inactivation (...) as the culture grows in density and the space widens, *the cells start to resemble pebbles.* This is the point when they should be "sieved" (...). *Gently* oscillate the plates (...), re-suspend the cells by shaking the tubes *up and down various times* (...). Dilute the cellular suspension in a *small quantity* of the culture.[8]

---

[8] I took this example from a stem cell university textbook, Lanza and Klimanskaya (2009, pp. 332–333, italics mine). On the importance of organised visualisation

Massimo Montanari notes how the practice of food preparation and that of medicines both make the same use of the word "recipe", from the Latin *recipere*, "to take" or "to chose", underlining in both cases the importance of the choice and combination of ingredients or active substances and of human intervention to compensate and balance taste and different properties (Montanari, 2009a, pp. 119–120).

In this case too, the comparison of the different practices has its roots in ancient tradition.

A favorite metaphor of some of the most influential authors at the origin of modern science to describe the research process, is that of game hunting, a *venatio* that leads to the exploration of hitherto unknown territories (Rossi, 1962, new ed. 2002, p. 61, 1991).

Francis Bacon compares research to "Pan's hunt", in reference to the myth whereby Pan succeeded where others had failed: during a hunt, he discovered the Goddess of agriculture Ceres in her hiding place. The experimenter is a sort of "hunter of nature's secrets" who proceeds, one experiment after another, "in the same way a hunter tracks his prey deliberately, step by step, guided by footprints and signs" (Eamon, 1994a, p. 283). His competence is of a highly practical nature, which contrasts with traditional philosophical reflection and requires "sagacity, or a scenting of nature" (Bacon, 1623, ed. 1857, I, p. 633, Eng. trans. 1901, p. 227).

Animals known for their predatory skills such as the lynx became the symbol of some of the first scientific academies (Accademia dei Lincei); a dog concentrated on pointing a hare became the emblem of the Accademia del Cacciatore established in Venice in 1596. The new experimenters were thus described as "hounds" sniffing for facts, instead of being bent over textbooks like their predecessors (Eamon, 1994a and b; cf. Cavazza, 1979).

Through these hunting metaphors, the new natural philosophy connected with one of the most familiar traditions of the courts in those times and therefore with some of its main interlocutors and patrons;

---

practices within various professional contexts, including that of scientific research, cf. Goodwin (2003). For an analysis of the relevance of actual knowledge in the field of science and technology, and for a comparison with the field of cookery, cf. Collins and Pinch (1974), Pinch (1997).

based on an argument technique that contemporary scholars of social representations would describe as "anchoring" (Moscovici 1961), natural philosophy thus defined and legitimised its new role based on its affinities with other practices that were already familiar and established.

For the physician and naturalist Francesco Redi (1626–1698), the metaphor of science as a hunt acquired a literal meaning, as he obtained specimens for his experiments from the many shoots of the Grand Duke of Tuscany; he was often given the role of "arbitrating the division between animals to be eaten and those on which to experiment" (Eamon, 1994b, p. 406).

One should therefore not be surprised that Redi was also in the habit of punctuating his experimental notebooks with observations on the culinary quality of the animals that he was studying and with suggestions on how to prepare them.

> Apicius and Atheneum would scold me if I forgot to mention this other observation — though it might not be relevant — that the Dolphin's brain is a very delicate food, just as good as that of a milk calf, or of any other animal eaten in the most lavish and creative kitchens; on the contrary, I would say, based on my experience, that it was much better, delicate and sweet. (Redi, *Osservazioni intorno agli animali viventi, che si trovano in altri animali viventi*, in *Opere*, 1809–1811, III, p. 389 — Redi, *Observations on living animals that are found in other living animals*, in *Opere* 1809–1811, III, p. 389)

More than one century later, the father of experimental physiology, Claude Bernard (1813–1878), compared the life sciences to a "superb and dazzlingly lit hall, which may be reached only by passing through a long and ghastly kitchen" (Bernard, 1865, Eng. trans. 1927).

At the beginning of the following century, a hospital magazine described the work amongst bacterial cultures, test tubes and inoculations of those working in the infectious diseases laboratories of St Mary's Hospital as that of "modern Mrs Beetons", thus comparing it to the hard working cooking practices inspired by one of the most widespread home economics manuals of the Victorian era, *Mrs Beeton's Book of Household Management* (1861), packed with scientific information as well as recipes (Waddington, 2010, p. 61).

To this day, a specialised journal like "Mycologist" regularly publishes a brief written column (*Cookery Corner*) beside experimental reports, in which the observed items (mushrooms) are at the centre of delicious recipes, such as that of "ink mushroom (referred to by its scientific name, *Coprinus Comatus*) stuffed with brie cheese".[9]

The importance of cooking references as a context for the cultural interpretation of science is confirmed by their satirical use. The ironic contrast between science and cooking has been used since the XVIIth century, to underline the distance that science was taking from common sense and daily life and to stigmatise the abstruseness of some debates and their indifference to practical problems.

One of the most famous examples of this is the poem called *La Crema Battuta* (Whipped Cream) by physician and poet Lorenzo Pignotti (1739–1812), in which a physicist has a subtle discussion with a metaphysician and a theologian about the properties of cream, until the cream itself speaks up to make fun of the vane bombastic nature of their hypotheses.

A serious Metaphysician, a solemn Theologian and a Physicist
were staring, their eyelash frozen
this work; but why the surprise?
Could it be that sciences abhor the pleasant smells of cooking?
No. rather that in their presence a chemical experience is taking place
The Physicist is focused on observing how, from time to time,
A little matter can expand to fill a vast space, and would courageously opine,
That the world consists of almost no matter at all.
That Nature is no more than a lightly whipped cream ball.
But suddenly, from within the cream
A mocking voice cuts short his dream:
"Reflect now about the image you have seen:
Where Matter is insignificant but a great wind is blowing through it,
Therein lies the true image of the vanity of human knowledge as we perceive it."
(Pignotti, 1823, pp. 103–105, ed. or. 1732)

---

[9] "Mycologist", May 1998, p. 71. The annotation at the end of the recipe ("Author's Copyright. No reproduction is allowed without authorisation"), introduces another

The "pleasant smell" of the kitchen is once again a positive reference, synonym of common sense and being down to earth, in opposition to "useless, vain" knowledge. On the other hand, the author's irony underlines and exacerbates science's tendency to keep its distance — with snobbish contempt ("abhor") even — from the sphere of cooking as the epitome of daily life and those dreary materialist practices from which it would be best to flee to assert its own status and distinctive prerogatives.

Not surprisingly later on, references to cooking would often be used during scientific debates to undermine the experimental results obtained by opponents or competitors. In these cases, "the knowledge of nature is compared to that of knowledge of the realm of art, politics or law" (Pinch and Collins, 1984, p. 522), but at times it is also degraded to the level of more mundane activities such as cooking. The reduction of science to cooking is a classic tactic when it comes to strategies to critically degrade scientific results.

Fig. 1.7 The smell of the geological "minestrone" by Charles Lyell attracts and dulls the wits of his colleagues (from Rudwick, 1975).

---

analogy that it might be interesting to explore, i.e. between recognition in science and recognition in gastronomy, and relative mechanisms of "intellectual property protection" in both areas.

In a series of satirical cartoons from 1831, British geologist Henry De La Beche mocks the theories by his rival Charles Lyell, who had introduced a new approach to geology with his *Principles of Geology*. Lyell is depicted with a chef's hat carrying a large cauldron of minestrone the smell of which induces the other scientists (all wearing thick glasses to underline their lack of clear vision) to follow him. The accusation made to Lyell seems to be that of having "cooked up" his results thanks to unclear and debatable methods. An accusation which Lyell may have inadvertently encouraged, having used culinary terms to describe his own work. In a private letter to Gideon Mantell, Lyell had in fact defined his "new general theory on climate" as a "cooking recipe" that made it possible to explain the climate at all latitudes as the result of the configuration of the earth and the sea, or based on a limited series of ingredients (Rudwick, 1975).

According to the mathematician Charles Babbage (1791–1871), "cooking" is one of the most common ways with which scientists can carry out fraudulent behaviors — for example by selecting solely empirical data that support their own theory: "Out of hundreds of observations, the cook must be very unlucky if he does not succeed in choosing at least fifteen or twenty of them to put on the table." Another typical "recipe" of the scientist-cook consists in applying different calculation formulas to discordant observations. For Babbage however, the scientist-cook runs great risks of being exposed in the long run, thus obtaining a "temporary reputation (...) at the detriment of long-term fame" (Babbage, 1830, pp. 174 ff, cf. also Merton, 1973).

Cold Fusion or Whipped Cream?

The case of cold fusion offers one of the most interesting examples of the similarities between science and culinary practices.

The case dates back to 1989 when American chemist Stanley Pons and his British colleague Martin Fleischmann announced that they had obtained low temperature nuclear fusion through an unpublished electrolytic procedure, using limited funding and experimental equipment. The discovery immediately became a huge sensation among specialists and the mass media around the world; American physicist Steven Jones claimed to have obtained the same result before the two chemists and over sixty laboratories worldwide announced that they had replicated the experiment with similar results. Nevertheless, the disparagement intensified and most of the physics community came to look very critically upon the experiment by Pons and Fleischmann.[10] A significant role in the case against and the demise of cold fusion was played by the "reduction of science to cookery", a recurring element in numerous comments, which also came from scientists.

> *What is brewing in that test tube* (title of one of the articles by Rubbia on "La Repubblica", 29 March 1989).

The fusion saga started when Pons and Fleischmann called a rushed press conference at the University of Utah in Salt Lake City, informing journalists that they had obtained cold nuclear fusion, *in a glass the size of a cocktail shaker* ("Corriere della Sera", 6 April 1989, my italics).

---

[10] For a more detailed reconstruction of events cf. Belloni (1989), Lewenstein (1992), Bucchi (1996).

This cold fusion... brings to mind the idea of a chocolate and vanilla ice cream cone... Apparently it is the greatest discovery of the century, or of the history of humanity even, which is why we like to imagine it in a so-called domestic context. Titanium? They sell it at the local supermarket! Deuterium? In Tuscany the farmers can't get rid of it fast enough! *There is a vague smell of ratatouille throughout the whole story, as if from one experiment to the next one could obtain cold fusion through the use of eggplants, carrots and a handful of basil.* ("La Stampa", 23 April 1989, italics original)

> One month ago, it was a mythical dream, and now everyone is preparing it as a humdrum risotto...that's understandable you'll say: if mothers jealously hand down to their daughters the formula for *Nocino* liqueur or for oven baked *Maccheroni*, why should we be surprised by this race for a patent that is worth various billions? ("Corriere della Sera", 21 April 1989)

Following the failure of many trials to replicate the experiment, a science journalist observed:

> Fusion is not like whipped cream, that whips up or not due to reasons unknown to the chef. ("La Stampa", 27 April 1989)

Pons and Fleischmann were even accused of having "re-heated an old soup", meaning that they had simply rehashed a similar discovery that had been announced during the first decades of the 20th century.

The two researchers also exposed themselves to the cooking metaphor when, during the first days of the events, they told journalists that they had started their experiments in Pons's kitchen.

> We were so enthusiastic at the idea that we decided to try it out in my kitchen. ("Il Sole 24 Ore", 25 March 1989)

> We threw ourselves into this adventure, starting immediately in my kitchen. ("La Repubblica", 25 March 1989)

> How is it possible that we obtained a result in our back shop, in our home kitchen in fact, that the best and highest paid brains in the world have been trying to obtain spasmodically at a cost of billions of dollars since the Fifties? ("L'Unita", 25 March 1989)

During an important meeting between researchers in Los Angeles, when scepticism was already high, Fleischmann explained his reluctance to provide full details of the experiment as follows: "A good chef never reveals all the ingredients he has used" ("La Repubblica", 11 May 1989). Among the many cartoons that were published about cold fusion, many referred to the kitchen, with fusion reactions exploding among the pots and pans of ignorant housewives and clumsy chefs.

Humorous devices such as caricature, parody and "unmasking" (in opposition to "travesty") contribute to what Freud calls *Herabsetzung*, or the degradation of an object or an individual.[11] Unmasking is particularly relevant in this case because it is used against someone "who has attached to himself dignity and authority which in reality should be taken from him"; "persons and objects who command authority and respect and who are exalted in some sense" to "degrade the dignity of

Fig. 1.8 Satirical cartoon on cold fusion (source: "Il Sole 24 Ore", 19 April 1989): "While she baked a cake for the bridge club, Mrs Emily Troodle discovered (…) FUSION!"

---

[11] Freud (1905, Eng. trans. 1917, pp. 325 ff).

individuals in that they call attention to one of the common human frailties, but particularly to the dependence of his mental functions upon physical needs".[12]

Thus the pretence of "purity" (i.e. having to do with science and truth that are considered untouchable by "human" elements such as interest, error and folly) by supporters of cold fusion is ridiculed and their presumed revolutionary object is scaled back and presented for example as the result of clumsy cooking.

According to Trevor Pinch,

> any scientific episode studied in detail, that reconstructs what scientists did on one particular occasion, retelling the story in a given way — that I call the "narrative of the clumsy scientist" whereby scientists appear to be incompetent and incapable of controlling their actions — destroys that piece of science. It becomes similar to daily life, subjected to the weaknesses of daily life.[13]

---

[12] Ibid. Garfinkel underlines the "irony between who the denounced person seems to be and the way he or she is really seen after the complaint" (1956, p. 422). Unfortunately, only very few sociological studies analyse humour and in particular humour related to science. A few exceptions: Mulkay and Gilbert (1982), Mulkay (1988). On the role of irony in scientific debate and criticism, cf. Hacking (1999), Becker (2011) and Bucchi (2011).

[13] Pinch (1992, p. 504).

In the Kitchen Science Becomes More "Human" (and Fun)

A cheese melting laser in a Swiss recipe for *raclette*; the nano-technological toaster; the aerodynamics of French fries; the effects of peanut butter on earth rotation.

These are just a few of the pieces of research published by the "Annals of Improbable Research", the journal of "research that makes people laugh and then makes them think", born from an initiative that dates back to the 1950s when virologist and scholar of scientific frauds Alexander Kohn, together with physicist Harry Lipkin, founded the "Journal of Irreproducible Results". With a mix of contributions that make use of forms and terminology that characterise science in a knowingly comical way and of authentic researches published in other "serious" journals, *Annals of Improbable Research* proposes to reveal "how very human and quirky and charming and downright enjoyable science is to the people who spend their lives doing it. Yes, scientists and doctors and science teachers are people, not inhuman geniuses" writes its founder Marc Abrahams, "Science is too human, too much fun, and too important not to laugh at" (Abrahams, 1998, p. 1). Since 1995, the journal awards the so-called Ig Nobel prizes to "results that cannot and should not be reproduced", bizarre researches that have been undertaken and published.

Thus science "relaxes", laughs and allows others to laugh about itself, revealing a more "human" side, one that is reassuring and approachable, that should encourage the general public to get closer to it. Using the same procedure described earlier in the case of cold fusion, but with completely different intentions, science is "downgraded" to the level of other daily life practices so that it takes on more familiar and reassuring connotations.

Within this strategy, cooking and practices linked to food take on a major role. Thus the Ig Nobel prize in 1996 for physics was awarded to Robert Matthews from Aton University in Birmingham, "for his studies of Murphy's law and especially for demonstrating toast often falls on the buttered side"; for chemistry to George Goble from Purdue University "for his blistering world record time for igniting a barbecue grill — three seconds, using charcoal and liquid oxygen"; for biology to Anders Baerheim and Hogne Sandvik from the University of Bergen in Norway for their work on the "Effect of Ale, Garlic, and Soured Cream on the Appetite of Leeches".

Some of these examples, however, demonstrate in a significant way the ambivalence with which science approaches cooking. On one hand, a proximity to cooking enables science to become more approachable by the general public and common sense; on the other hand, this leads to science receding from them: research on buttered toast or the effects of soured cream and leeches are examples of ridiculous and useless science and thus deserve the Ig Nobel award.

This aspect is all the more obvious in interviews to famous scientists, including "real" Nobel prize winners, hosted by the journal, and to whom questions on cookery and food are often asked, as was the case with the 1989 Nobel prize for chemistry Sidney Altman:

> *How many chips per glass of beer are optimal?*
> Well, that question is typical of a non-scientist. There is no regard whatsoever to quantitative observation or to maters of scale. What kinds of glass are we talking about? What sizes? What sizes of chip are we talking about? What density of chip? (...) this is a matter of individual choice and has nothing to do with science. (Ibid., pp. 29–30)

The reference to cooking as a humoristic device also characterises sophisticated parodies about science such as those by the French writer Georges Perec. In his *Experimental Demonstration of the Tomatotopic Organization in the Soprano (Cantatrix Sopranica L.)* Perec replicates precisely the format and conventions of a scientific article (bibliography, graphics, methods and materials description, results) where he describes an experiment on the effect that throwing tomatoes has on the volume of a soprano lyrical singer. In the paragraph on *Histology*, he notes that:

> At the end of the experiment, the sopranos are subjected to a perfusion of olive oil and Glenn-fiddish (sic!) at 10%, and placed in incubation in orange juice at 15% for 47 hours. Uncontaminated frozen 2 cm sections are fixed in a ∂-strawberry sorbet and observed under light and heavy microscopes. (Perec, 1996, Eng. trans. 1996, p. 120)

Among the bibliographic references at the end of the article are a series of contributions such as *Seventeen easy cabbage and garlic recipes; I. With Tomatoes* by a certain O. Chou (homophone of *Chou* or Cabbage) and A. Lai (*à l'ail*, with garlic); *Biological effects of ketchup splotching* by a certain Heinz (a famous ketchup brand); and a contribution by Mace & Doyne (homophone of *macédoine*, "a mix of cubed vegetables at times a synonym of "Russian salad").

From Science *à la carte* to *Take-away* Science?
Medicine and cookery according to Socrates

So what does the metaphor of cookery as a science tell us — and vice versa the metaphor of science as cookery — about the relationships between science and society and their transformations?

In Plato's *Gorgias*, Socrates repeatedly opposes cookery to medicine. Cooking, he claims, is to medicine what rhetoric is to justice. It flatters the body and fools individuals by offering short-term pleasures instead of a more substantial wellbeing.

> Cookery simulates the disguise of medicine and pretends to know what food is best for the body; and if the physician and the cook had to enter into a competition in *which children were the judges, or men who had no more sense than children*, as to which of them best understands the goodness or badness of food, the physician would be starved to death. (Plato, Gorgias, 464e, italics original)

"I shall be tried" Socrates concludes, referring to the possibility of being prosecuted, "as a physician would be tried in a court of little boys at the indictment of the cook. What might he reply under such circumstances, if someone were to accuse him, saying, "O my boys, many evil things has this man done to you: he is the death of you, especially of the younger ones among you, cutting and burning and starving and suffocating you, until you know not what to do; he gives you the bitterest potions, and compels you to hunger and thirst". What do you suppose that the physician would be able to reply when he found himself in such a predicament? If he told the truth he could only say, "All these evil things, my boys, I did for your health," and then would there not just be a clamour among a jury like that? (Ibid., 521 and 522b)

For Socrates' argument, it is crucial to represent the public as being immature: only a poorly educated child-like public would prefer a mundane activity such as cooking or gastronomy to its scientific counterpart, medicine.[14]

The *Superquark* section *Science in the Kitchen* does not oppose cooking and science so starkly. In times of increasing pressure to engage the public, science cannot limit itself to looking down at common sense. The question therefore is not that of choosing one option (science) and eliminating the other (cookery), but rather of inserting science into cooking by incorporating areas of common sense within a scientific perspective, in order for cooking to become a subdomain of science. This is a very common form of expanding the boundaries for contemporary science — consider for example how some sectors of biology or more recently neuroscience have extended or enhanced their competence in aspects that used to be traditionally the prerogative of philosophical reflection, religious faith or political debate. From a more internal point of view, within science itself, since the middle of the last century, techniques and concepts from physics have entered the sphere of biology, to the point that some scholars consider it typical of science to develop by extending itself and its languages towards ever more new territories.[15]

What has remained consistent, from Socrates to *Superquark*, is the representation of the public as being mainly passive: there a jury of children, here a cooking class of women — historically a key target of scientific popularisation since the success of Algarotti's *Newtonianism for Ladies* (1727) and Lalande's *Astronomie des dames* (1785), described as "symbols of ignorance, good will and curiosity" (Raichvarg and Jacques, 1991, p. 39).

Nevertheless, whilst in Socrates' vision science can still remain on a pedestal, until the public becomes sufficiently cultivated, enlightened and mature enough to be able to climb up and reach it, in the television programme we see scientists ready to descend into the kitchen without hesitation and to dirty their hands among pots and pans.

---

[14] Cf. Latour (1997).

[15] Cf. Cameron and Edge (1979), Mulkay (1974), Bucchi (2010a).

31

According to historians, the XIXth century explosion of scientific popularisation can be interpreted as the evolution of a new model of large-scale popularisation, from the previous communication model that targeted a highly selected and motivated public. Thus, a *science du chef* — with its fixed menu, whose delicacies the general public was invited to taste — next to a more traditional *science à la carte*. Could it be that the day has come, for the public, to try out a science that is more common, affordable and ready to use; a *take-away* science of sorts?

MAIN COURSE
The Science of Chicken

Does he cook in the kitchen yet?
No. But he uses the kitchen for many other things.

Kitchen Stories

To Die for Science: Bacon's Chicken

Snow continues to fall on London during the first days of Spring 1626. But Francis Bacon, the Viscount of St Albans, great theoretician and apologist of newborn modern science, does not fear bad weather. One of his habits is taking a walk in the woods in the company of young Thomas Hobbes, whose task is to write down the Viscount's intuitions. When it rains, Bacon often requests the carriage top be left open, "to receive the benefit of irrigation, which he was wont to say was very wholesome because of the Nitre in the aire and the Universall Spirit of the World" (Aubrey, 1950, p. 134, original italics).

On that day in 1626, Bacon is travelling by carriage towards Highgate with his friend Witherborne, King James's doctor. His mood is not the lightest. A few years earlier, Bacon has been condemned on corruption charges for conceding monopolies to the jewelry trade during his time as Lord Chancellor. He has served his sentence with a 40,000 sterling fine, a ban from public office and even a few days imprisonment in the Tower of London.

Leaning out of the carriage window, Bacon invites Witherborne to note that as the wheels of the carriage roll over the snow, they reveal patches of bright green grass; grass which looks as if it has freshly grown. Bacon attributes the phenomenon to the snow and ice, and puts forward the hypothesis that these might possibly be used to preserve fresh foods, just as one does with salt.

Witherborne replies that the idea seems absurd to him. At that, an irritated Bacon asks the coachman to stop immediately. He gets out of the carriage, oblivious to the outside temperature, and sets off on foot until he finds a house which he thinks might have what he needs. He knocks on the door and asks the woman who appears if he can buy a chicken, requesting that she gut it for him there and then. Grabbing

35

the chicken from the woman's hands, the Viscount then bends over and stuffs snow into the chicken with his bare hands, then puts it into a bag and covers it with snow. A few minutes later he starts to feel unwell because of the cold. Witherborne immediately realises that there is no time to waste and instead of taking him home, takes him to the Count of Arundel's who lives nearby, and puts him straight to bed. But the bed is damp, his condition deteriorates rapidly, and a few days later, on April 9, 1626, Bacon dies at the age of 65, of "colde and suffocation" according to his friend Hobbes, of acute pneumonia according to others.

The episode of Bacon's chicken, of which the authenticity is still debated[16] by historians, soon became one of the founding myths of the scientific revolution such as the apple which fell on Newton's head — the epitome of sudden intuition and serendipity[17] — or Copernicus hugging the first copy of his *De revolutionibus orbium caelestium*, the proto-type of unselfishness and dedication to scientific progress: "the great apostle of experimental philosophy was destined to become its martyr", commented the historian Thomas Macaulay.[18]

It is an admirable incarnation — literally — of the stubbornness with which the new science intended to tackle empirical data; of the absolute dedication with which scientists studied nature (to the point of risking their lives). The frozen chicken experiment is often the only anecdote for which the general public remembers the great philosopher and politician.

It does not matter much, of course, that until that day Bacon had actually never undertaken any experiment, and that his playing around with snow and chicken did not actually feature many of the elements which should actually characterise an experimental trial (for example, he did not have a "control chicken", i.e. non-refrigerated, or preserved in

---

[16] Cf. for example, Jardine and Stewart (1998), Gribbin (2007).

[17] The expression Serendipity refers to the accidental nature of some scientific discoveries. Cf. ch. 2, par. 3.

[18] Thomas Babington Macaulay, "Lord Bacon" (1837), in *Critical and Historical Essays contributed to the Edinburgh Review*, London, Longman, 1949, p. 372. A destiny of which Bacon had probably already had an intuition, when, in a letter dated just prior to his death, he compared himself to "Plinius the Elder, who lost his life trying an experiment about the burning of the mountain.

salt); nor, more fundamentally, that no one had ever provided feedback on his final result (was the chicken preserved? For how many days? Was it then eaten? And if so, what were the effects?).

The fact that this episode has become part of our collective memory and popular culture is proven by its many references in the most diverse contexts, from international law controversies to poetry.[19] And although the experiment ended tragically, it also inspired humour. One famous example is Woody Allen's story on the invention of the sandwich, invented by the homonymous Earl of Sandwich, a figure who seems to resemble Bacon in many ways, and whose adventures are a hilarious intertwining of science and cooking.[20]

The story's main character is the son of the King's Chief Farrier whose essay on *The Analysis and Attendant Phenomena of Snacks* arouses the interest of his schoolteachers. The Earl is expelled from Cambridge "charged with stealing loaves of bread and performing unnatural experiments on them".

> His first completed work — a slice of bread, a slice of bread on top of that, and a slice of turkey on top of both — fails miserably. Bitterly disappointed, he returns to his studio and begins again. (...) He exhibits before his peers two slices of turkey with a slice of bread in the middle. His work is rejected by all but David Hume. (...) Destitute, he can no longer afford to work in roast beef or turkey and switches to ham, which is cheaper. In the spring, he exhibits and demonstrates three consecutive slices of ham stacked on one another; this arouses some interest, mostly in intellectual circles, but the general public remains unmoved. Three slices of bread on top of one another add to his reputation, and while a mature style is not yet evident, he is sent for by Voltaire. (...) He works day and night, tearing up hundreds of

---

[19] Cf. e.g. J. Baccus (2004). The poem is by Pip Wilson: "Against cold meats was he insured? For frozen chickens he procured — brought on the illness he endured, and never was this Bacon cured".

[20] W. Allen (1972, 1991, pp. 178–181). From an anthropological point of view, it is interesting to note the existence of an urban legend according to which the ghost of Bacon's chicken appears occasionally to this day in the area of Highgate. http://news.bbc.co.uk/dna/place-lancashire/plain/A14042099

blueprints, but finally — at 4:17 A.M., April 27, 1758 — he creates a work consisting of several strips of ham enclosed, top and bottom, by two slices of rye bread. (...) At his funeral, the great German poet Holderlin sums up his achievements with undisguised reverence: "He freed mankind from the hot lunch. We owe him so much." (Allen, 1972, 1991, pp. 178–181)

Just like the comic strips on cold fusion, the research presented by the Annals of Improbable Research, and the Ig Nobel prizes (see chap. *Starter*), comic effect is here produced — among other things — by the short circuit between the intellectual practices of science and more mundane food preparation. The "invention" of a common food known to all as the sandwich is described following the typical steps of scientific research and invention: failures and incomprehension, cumulative growth of knowledge, sudden intuition, and final recognition from colleagues and political powers. By doing so, the status of scientific research is thus comically "downgraded" to a daily practice, whilst gastronomic activity is ridiculously upheld to the status of intellectual conquest.

The Enlightenment Philosophers' Chicken

Poulet, Poule, Poularde (Dietetics and Medical material) Chicken, Hen, *Poularde.* An old hen makes for an excellent broth when boiled with other meats for soups, and even when she is fat, her boiled meat is of quite agreeable taste and excellent for the health: it is especially good for those who are convalescing (...). The use of chicken as medicine, or as a medicinal food is as well-known as its dietetic properties; it ordinarily becomes part of cooling and soothing broths accompanied with herbs of equal properties, and flour based seeds etc... It is a mistake which even reputable doctors fall for, that of stuffing cold seeds into chickens destined for this purpose; because cutting seeds don't "give" anything during the cooking process (...), we are in the habit of taking off its skin when preparing broth and chicken water; this practice is quite useless. (*Encyclopédie*, pp. 202–203)

These are the cooking and therapeutic instructions in one of the entries dedicated to chicken in the *Encyclopédie* (1751–1781) by Diderot and D'Alembert. A theme emerges here that will become recurrent in the intersections between science and cookery: common sense laid bare and disdained, widespread cooking practices deemed naïve and unaware.

Chicken comes up in many of the *Encyclopdie's* entries, as a source of meat that is beneficial for "delicate stomachs", indicated for various uses including therapeutic ones, and suitable to be farmed on a wide scale. Detailed drawings and diagrams illustrate the structures that are optimal to raising the animals and introduce the "art of egg hatching" with adequate incubator ovens.

The introduction of said ovens — originally from Egypt — and new breeds, especially from China, contributed to trigger during those years a "chicken breeding-craze comparable to the Dutch tulip mania of the 17th century" (McGee, 1984, p. 71).

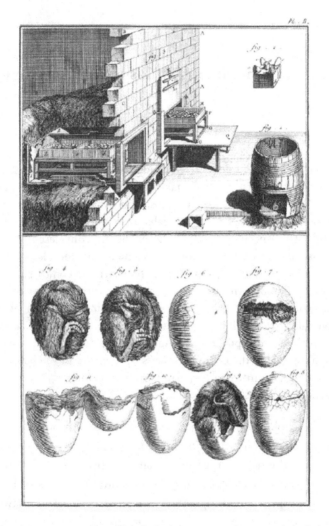

Fig. 2.1 *Rural Economy — How to hatch chicken eggs. Incubators and chick development,* from: *L'Encyclopédie by Diderot and D'Alembert,* Plates vol. 1, 1762.

These innovative poultry farming methods had already been compared with an interesting analogy to metal transmutation by authors during the previous century such as physician Giovanni Imperiale in his *Le notti beriche, overo de' quesiti e discorsi fisici, medici, politici, historici e sacri* (1663) (*The nights in Monte Berico, or of questions and discussions on physics, medicine, politics, history and the sacred*) if one can obtain a chicken from an egg in little time and thanks to heat, why could one not obtain gold in the same way from other metals?

> (...) if through the artifice of heat it is possible to generate animal bodies, that are much more perfect than those of metals, it should be all the more possible to use the same method to produce those that are far less perfect. It is already well known that in Egypt, chickens were born thanks to the proportionate heat from eggs placed in ovens, which is why they are abundant in that country. Thus gold could be born from metal, just as chickens are from eggs. (Imperiale, 1663, p. 8)

The new XVIIIth century taste moves away from "heavy, greasy meats" and from "strong tasting game" preferring chickens and birds (Camporesi, 1990; new ed. 1998, p. 51).

René-Antoine Ferchault de Réaumur (1683–1757), one of the greatest naturalists of those times, member of the Académie des Sciences, dedicates a treaty in several volumes on the rational organisation and the most efficient technology to raise and farm chickens and other forms of poultry. The treaty is *Art de faire eclorre en toute saison des oiseaux domestiques* (*The Art of making domestic birds hatch in all seasons* 1749, trad. Eng. 1750). Among the themes of the various chapters, are instructions on how to build ovens to make the eggs hatch, on the ways "of hatching chickens without having the eggs sat on by hens", "the different foods that may be given them". Finally, there is a chapter dedicated to the "philosophical amusements" that chickens can potentially offer. Here Réaumur invites colleagues, housewives and curious people to make note of the quantity and quality of their produced eggs, in relation to the different types of grain they are fed; and to experiment with the eggs, by greasing the shells with butter, fat or olive oil to verify if in so doing they keep for longer. To those who objected, considering these experiments "of poor relevance", Réaumur replied that they only had to note the diffusion and the broad use of eggs and chicken in our diets.

Therefore, science should neither hesitate to enter the kitchen nor the barnyard. Rather modern is also the idea is that the social status of naturalists could benefit from recognised practical utility of their knowledge. Domestic farming animal *par excellence*, the chicken is considered by Réaumur to be the ideal object for applying reason and experimental methods; a living organism that lends itself to an almost mechanical cause and effect analysis, in which the output — the quantity and quality of the eggs and the meat — is directly correlated to the input — fodder and farming methods — and the animal is a sort of machine whose performances can be maximised.

The pages in which the scientist wonders about the possibility of obtaining ready aromatised and spiced meats by inducing the animals to feed on ingredients that are used in the preparation of foods (such as cinnamon or nutmeg) are particularly significant.

> "I have not as yet attempted any of these experiments, but I have one to offer that was made by mere chance, and which gives very good hopes for the rest. Mr Bouvard of the academy of sciences, told me that a couple of turkeys having once stolen into a garden of his father's, that was not over and above well-guarded, and where there was a border of onions, they found the green part of them very palateable; they were left to feed on them for some days, after which they were killed. Their flesh was found of an exquisite taste, which had never been perceived in that of turkeys, a flavour like that of pheasants and venison: their flesh had been intimately seasoned by the onion. People say, on the other hand, that the turkeys that pass, in their way to Paris, through the forest of Fontainebleau, and make a stop there, have a flesh which the juniper berries they eat upon the road have rendered very unpleasant to the taste. The above-mentioned experiment of Mr Bouvard, is then an invitation to us to feed not only turkeys, but also chickens, with onions, leaks and garlic." (Réaumur, 1749, II, pp. 271–272, trad. Eng. 1750, p. 424, italics mine)

This is how the new knowledge and scientific methods insinuate and incorporate traditional competences and knowledge into food preparation. The entry of the *Encyclopédie* also warned against common "errors" and practices that when scrutinised were revealed to be "useless" and unfounded. "Although people are very well skilled in the art (...) of procuring fat fowls, capons, turkeys goslings and geese, (...) although it be a part of the business of the poulterer, (...) and a prodigious multitude of these fattened things (are) yearly conveyed to Paris, there remain very probably a great many things to be known with regard to the most speedy method of fattening them, with regard to the cheapest way of bringing it about, with regard to the practice that procures both the fat and the flesh of the nicest quality" (Ibid., p. 425).

Chicken and Children First:
Pasteur's chickens, or how discoveries sometimes happen when scientists are on holiday

Three animals and a child mark the rise to scientific glory and the notoriety of Louis Pasteur.

A sheep, or rather, fifty sheep are at the centre of the famous public experiment during which Pasteur tested the efficiency of his anthrax vaccine, thus challenging the scepticism of many of his colleagues. The resonance of his experiment was such that the newspaper The *Times* sent a special correspondent from London to follow it, invited by Pasteur himself (a clever communicator as much as he was a genius researcher). The chronicles of those times mention a festive crowd who welcomed Pasteur on his arrival at the train station when he finally came to check the results, celebrated for weeks by the popular press that contributed to forge his reputation as a great scientist and national pride.[21]

The second animal is a rabid dog, the one who bit the little Joseph Meister. Pasteur injected him with the vaccine that he was experimenting on, and three months later he was able to inform the Académie des Sciences that the boy was in good health. Meister remained close to him and became the custodian of the Pasteur Institute until 1940, when he chose to commit suicide rather than witness the profanation by the Nazis of the crypt where his saviour was laid to rest.

The third animal is actually the first one from a chronological point of view, which paved the way for a series of successes — often accompanied by much debated experiments and animated controversies with his colleagues — in the field of vaccinations.

---

[21] For a more detailed reconstruction, see Bucchi (1997; 1998a).

It is the summer of 1879 and Pasteur is exhausted. His repeated trials to solve the problem of the so-called "chicken-cholera" have led to no significant results. The disease has been known for about ten years, since an Alsatian vet observed "granulations" in animal bodies, after an entire hen house had been exterminated in just one day. It is Toussaint, a professor at the Veterinary School of Toulouse who discovers that a microbe is responsible. Toussaint decides to send to Pasteur the head of a cock that has died of cholera, a gesture that seems macabre but actually signals a great open-mindedness for the sharing of scientific work. Pasteur is struck by the extreme virulence of the microbe, so that "it only takes the smallest drop of culture on a few breadcrumbs to kill the chickens". His communications to the Académie des Sciences and the Académie de Médecine describe the suffering of the poultry in extremely vivid terms.

> The animal suffering from this disease is powerless, staggering, its wings droop and its bristling feathers give it the shape of a ball; an irresistible somnolence overpowers it. If its eyes are made to open, it seems to awake from a deep sleep, and death frequently supervenes after a dumb agony, before the animal has stirred from its place (...). (Pasteur, in Vallery-Radot, 1900; Eng. trans. 1911, p. 298)

For Pasteur, the practical importance of this new challenge is no less obvious than its scientific relevance. As had already happened with his studies on wine and on the silk worm — and as was to happen with those on the anthrax vaccine — Pasteur was called to intervene on terrains that were vital for agriculture and farming, and thus for French economy and politics. Like wine, chicken is one of the symbols of the French culinary culture, and the historical and political context of those years was ever more sensitive to patriotic and nationalistic callings. Pasteur would go on to prove many times that he was savvy in making the most of this sensitivity to enhance his researches by presenting them as a symbol of France's glory and superiority — especially compared to their neighbouring Germany. Pasteur would often use nationalistic arguments for example, to criticise theories by the spontaneous genera-tion and the works of Koch for being "German", and went on to argue, in his correspondence with Napoleon III for "the necessity to maintain the French scientific superiority on that of rival nations" (Pasteur, 1939, VI, pp. 10–11; Bucchi, 1998a).

But let us go back to the summer of 1879. The work on chicken cholera has stalled, and Pasteur has decided to retreat as was his habit to his family home in Arbois. Before leaving, he gives instructions to his assistants to continue "cultivating the germ at regular intervals every twenty-four hours" (Geison, 1995, p. 40; Vallery-Radot, 1900 and 1931, p. 427). But it is very hot, there is much to do in the laboratory, the chicken cholera appears to be a dead end, and the assistants forget the cultures on their shelf. At his return, Pasteur picks up one of the old forgotten cultures and tries to use it to infect an animal. Nothing comes of it. He then tries to infect the same "resistant" animal with a fresh, and undoubtedly virulent culture. Still nothing, the chicken remains healthy. Pasteur remains silent for what seems like an eternity to his collaborators, then starts to shout as if he had just had a vision: "Don't you see that these animals have been vaccinated!" (Dubos, 1960, p. 114). The attenuation of the pathogenic agent that Pasteur will later attribute to the effect of oxygen on the forgotten cultures, has made them immune to the illness. French chickens are safe!

Just like Bacon's chicken, the event would take on an almost mythological trait and become one of those episodes that would lead Pasteur to assert, on numerous public occasions that "Chance favours only the prepared mind". Many studies and theories would interpret this as one of the most momentous examples of so-called *serendipity*, an expression "originally coined to mean a quality of the actor in a happy accidental discovery, it has with use become coterminous with the whole event of accidental discovery, and even with the object of such a discovery" (Merton and Barber 2004, p. 102).

Detailed historical studies have actually documented a different version of events: it was neither a coincidence nor the assistants' distraction in forgetting to revive the cultures thus involuntarily producing its attenuation, but a precise strategy of one of Pasteur's most important collaborators, Emile Roux, who took advantage of Pasteur's absence to carry it out (cf. Cadeddu, 1985).

The confirmation seems to come from Pasteur's reticence over the following months to provide an effective explanation of the phenomenon; a reticence that might have reflected his incomplete understanding of the vaccine and its implementation method, and that would attract much criticism from his colleagues. A year after the triumphant experiment, Lord Joseph Lister, a pioneer in antisepsis in the field of surgery asked him if there was an available publication on the topic. Pasteur answered:

> I would gladly tell you everything I know on the attenuation of very infectious viruses; but I really cannot, not for a futile desire to hide a secret, but because (...) I am not satisfied with my findings (...) I would like to obtain other attenuated viruses. If I did I think we could shed light on a few unclear points that imprison me in a silence that is more painful for me than for the public. If you have the time, inoculate a few chickens. (Pasteur in Cadeddu, Ibid., p. 104)

The Chicken that Newton Did Not Eat:
The scientist's ascetic body

Dinner had been served for a while, but Isaac Newton was immersed in his studies and had not yet appeared in the dining room. His friend William Stukely was growing famished, and more and more impatient. He eventually lifted the lid off the dish where he discovered a chicken. He ate all of it then furtively placed the lid back on. When Newton finally turned up, he greeted his friend and sat down to eat. He lifted the lid, and saw that only the chicken bones were left. Then he coolly commented: "How absent we philosophers are. I really thought that I had not dined."[22]

This is another way in which science relates to cooking: by snubbing it as if it were a mere distraction, unworthy of competing with research activities. The scientist, by focussing on his intellectual mission, emancipates himself from material needs such as food. This quasi-ascetic characterisation of the figure of the scientist, his indifference for mundane practices such as eating or sleeping has been a recurring trait since the origins of modern science: Bacon's mother, Lady Ann Bacon, used to attribute her son's digestive problems to his habit of going to bed late, and staying awake to "meditate on who knows what instead of sleeping" (cit. in Gribbin, 2007, p. 52).

Hobbes' biographers describe his diet as "moderate… had his breakfast of Bread and Butter (…) he never was, nor could not endure to be, habitually a good fellow, i.e. to drink every day wine with company, which, though not to drunkenness, spoils the brain" (Aubrey, 1898, p. 350). In Boyle's eulogy it is claimed that had neither eaten nor drunk

---

[22] Brewster, *Life of Newton*, 1831, cit. in Shapin (1998, p. 21). The event appears in many popular texts, see for example Sagredo (1960).

for over thirty years "to gratify the variety of his appetite, but only to sustain his nature" (Burnet, cit. in Shapin, 1998, p. 38). In almost identical terms, the biography signed by Adrien Baillet (1691) ponders on the sobriety and moderation of Descartes who was "capable of not drinking a drop for a month".[23]

Numerous stories and anecdotes underline Newton's indifference to food; he would "leave his meals waiting for two hours", "often ate his cold dinner for breakfast" and "fattened his cat with all the food he left on his tray".[24]

The frugality of the Scottish chemist and physicist Henry Cavendish was memorable; not only did he avoid all direct contact with his staff about dietary requirements — he used to order his meagre meals in writing — but he regularly invited guests to his table, including his colleagues from the Royal Society and offered them "nothing but a leg of mutton" (cit. in Shapin, 1998, p. 41).

What follows is the report of the crucial days that led Pasteur to his discovery of fermentation (see chap. *Drinks on the Side*), provided by a very successful popular biography, a significant indicator in popular imagination of the scientist and the research process that figures such as Pasteur contributed to shape.

> He *ate*, slept and dreamed (...) with his ferments by him. They were his life (...). He waited and signed some vouchers and lectured to students and came back to peer (...) at his precious bottles and advised farmers about their (...) fertilizers and *bolted absent-minded meals*. (De Kruif, 1926, 1954, pp. 66–67, italics mine)

The decisive moment of his intuition of the fermentation process is narrated as follows:

> He put a drop from the bottle before his microscope. Eureka? The field of the lens swarmed and vibrated with shimmying millions of the tiny rods. "They multiply! They are alive!" he whispered

---

[23] Cited in Shapin (2000, p. 145).

[24] More, *Isaac Newton* (1934); Westfall, *Never at Rest* (1980), cit. in Shapin (1998, p. 41).

to himself, then shouted: "Yes, I'll be up in a little while!" to Madame Pasteur who had called down begging him to come up for dinner (…). For hours he did not come. (Ibid., p. 65, italics mine)

The one person who epitomises this contrast with the context of daily life in this case, is Mrs Pasteur — once again a woman (cf. chap. *Starter*) — who prepares delicious meals in vain that are destined to get cold while her husband proceeds uninterrupted with his research.

More generally, the lack of interest in food, and the frugality and sobriety of their eating and drinking habits have contributed to forging the public iconography of the scientist. It is an iconography that emphasises the intellectual and disincarnated character of the man of science, his detachment from material needs and even from a bodily dimension, as shown by the last images of Darwin, where the only things left for the public to see are "his beard, his hat and his eyes" or the recurring images of his by then empty study where "all that was left was the intense impression of his mind (…). Darwin, as a physical presence, had almost disappeared (…) his intellect appeared to the public in an almost completely disembodied form" (Browne, 1998, pp. 279–280).

This iconography has common traits with the way society and culture recognise certain figures as "depositaries of truth and value, be it religious, scientific, philosophical or artistic" within "virtuous and sacred conceptions of knowledge attributed to special persons and bodies" (Shapin, 1998, pp. 43–45).

There is no doubt, however, that one of the most influential models from this point of view has been that of religious iconography, that has always considered the disinterest for the pleasures of eating, and even "superb abstinences", as a fundamental characteristic of some of its most illustrious exponents. Doctors and natural philosophers of the XVIIth century used to suggest furthermore that the blockage of skin pores, which hermetically and no less symbolically sealed the body from external factors, was key to explaining the capacity of some ascetics to survive long periods of fasting.[25]

---

[25] Camporesi (1994); Imperiale (1663). Boyle also studied the matter in his *Tentamen Porologicum* (1686).

In biographies like that by Hobbes and Boyle, modern culture "ingeniously re-elaborated" (Shapin, 1998, p. 36) the Christian precepts of moderation and soberness on a dietary level too, and Baillet, who was a well-known writer of the life of the saints, was chosen as Descartes' biographer by the philosopher's disciples.[26]

The public iconography of scientists, as discussed in the above-mentioned examples, developed over the next few centuries, and particularly during the XIXth century, together with the establishment of science from a social and institutional standpoint and with the expectation by emerging civil society in many countries, that lay figures of high symbolic value should be available beside religious saints. It is certainly significant that during those same years, an increasing number of portraits, biographies and monuments dedicated to inventors and scientists began to appear. One particularly famous statue is that of Faraday in the atrium of the Royal Institution (1876); as for Italy, one should recall among others the monuments dedicated to Volta (1878) and Galvani (1879).

During the following century, the characterisation of scientists' "exceptionality" continued, albeit with inevitable transformations, to contrast with the needs and material aspirations of the common man, which were often symbolised by references linked to diet. Albert Einstein, perhaps the most relevant and celebrated icon in the field of science of the 20th century, wrote the following in his autobiographical notes:

---

[26] Cf., for example, Irace (2003).

When I was a fairly precocious young man I became thoroughly impressed with the futility of hopes and strivings that chase most men restlessly through life. (...) *By the mere existence of his stomach* everyone was condemned to participate in that chase. *The stomach might well be satisfied* by such participation, but not man insofar as he is a thinking and feeling being. (A. Einstein, *Autobiographical Notes*, in Schlipp, 1958, pp. 3–4, italics mine)[27]

---

[27] Naturally this tradition does not exclude alternative discourses such as those associated for example with figures of scientists who were more "libertine" or sociable such as Richard Feynman of Kary Mullis. These can seemingly represent ways of "role distance" in a Goffmanian sense, a playful individual distancing from a role model that as such re-establishes the role's general codification (Goffman, 1961). A broader discussion should also take place around the theme of an eventual "de-sacralisation" of the public figure of the scientist in our times and of the consequences this has on the status of scientific expertise in contemporary society. The impact of James Watson's famous autobiography (1968) is at times considered a symbolic milestone, having clarified to the general public that "scientists too are human [..] all too human" (Merton, 1973, p. 325). Cf., for example, Shapin (2008). For a detailed analysis of Nobel laureates as civic saints of science, see Bucchi (2017, 2018).

Eating Chicken before an Audience:
Science in public and the body of the leader

In 2006, the threat of bird flu is raging in the press and in public opinion. Among the many images through which the disease dominates screens and pages, one is quite peculiar. It is the image of politicians busy biting into a chicken — generally roasted or on a skewer — in an effort to reassure public opinion on the capacity of institutions to keep the spread of the virus under control. Among those portrayed eating a chicken in front of cameras and reporters are Italy's Minister for Agriculture Alemanno and Turkish prime minister Erdogan (when the first cases of infection emerged in Turkey).

These "public demonstrations" are important for at least two reasons.

The first is that they signal a contamination between the styles of communication of politics and those typical of science. A contamination that has now become a regular feature especially when it comes to issues of great public impact linked to health. During a press conference, at the time of the "mad cow" disease, the English Minister for Agriculture John Gummer tried to reassure public opinion by offering a hamburger to his daughter Cordelia (whose namesake is that of King Lear's daughter as one ironic commentator noted...). And although only a handful of American journalists were able to attend the shocking moment when a doctor, sceptical about the official causes of AIDS, injected himself with the blood of an HIV positive patient, almost everyone in Italy still remembers the moment when the immunologist Fernando Aiuti kissed a girl infected by the virus (Bucchi, 1998b).

Naturally one could object — and some have done so — that the scientific value of these demonstrations is rather dubious. Who can guarantee for example, that the hamburger eaten by the English minister's daughter was really made of beef?

Public demonstration has, however, been widely used by scientists themselves. As mentioned earlier, Pasteur himself resorted to a public experiment in order to convince his colleagues who were skeptical about the efficiency of his anthrax vaccine. A memorable simple "experiment" was the one with which the physicist and Nobel Prize winner Richard Feynman "demonstrated" in front of television cameras the causes of the explosion of the Challenger in 1986 (see chap. *Drinks on the Side*).

As a matter of fact, scientific experiments started as public demonstrations: Boyle did his historical experiments with the air pump (Shapin and Schaffer, 1985) in front of a group of esteemed gentlemen (women were not admitted) at the Royal Society. Historian of science Simon Schaffer retraces their origins to the so-called "assays" in which, with instruments and often elaborate "social scenography", the resistance, the purity or the possible adulterations of materials, coins, therapeutic remedies, clothes, food and wine were tested in modern times. Some assays took on the role of proper "*disfide*, courtly contexts", to try and solve intense controversies among the supporters of different theories or test the efficiency of experimental apparatus by copying other models of courtly challenges linked to friendly rituals such as the so-called "*cimenti della tavola*, postprandial trials of virtue".

In 1660, the above-mentioned Redi was involved in a long elaborate assay put together upon the request of the Grand Duke Ferdinando II de' Medici to test the efficiency of a laxative and thus evaluate if it might be used for courtly consumption. The product was experimented on about fifty subjects forcibly recruited among the citizens of Florence and segregated in order to monitor the effects of its administration (Schaffer, 2005).

Redi offers an extraordinary example of an almost literal conjunction between an "experimental assay" and a culinary "taste sampling". Having at hand some deer brains in order to "observe their fabric", he decided to challenge the widespread belief that

> Deer brain was a disgusting thing, almost repelling to eat, and extremely harmful to human health; thus there was not a single gentleman at Court, who out of courtesy or fear had risked putting deer's brains on his table; However (...) because these brains seemed to me to be chubby, beautiful, healthy and

substantial, I took the risk, notwithstanding the fact that my servant was embarrassed to bring this Lutheran roguery into the kitchen (…), to have a whole pot-full fried in virgin lard and once it appeared, warm and golden on the table, I admit I gobbled almost all of it down, and found that through this repeated, true and secure experiment, deer's brains are an agreeable food, very tasty and healthy and much better that the brain of a pig or a veal, not to mention that of a dolphin, in fact in my view it is better than all other brains because it can be eaten during Lent and the eve of high holidays.[28]

In the same way contemporary politicians publicly bite into chicken, Redi challenges in first person, as a scientist and as a dinner guest, diffused and — for him — prejudicial preoccupations.

But what is most notable is how in this case science and cooking are combined not only from a cognitive point of view but also from an aesthetic one. To Redi, the deer brains look beautiful and well structured, healthy and substantial, thus they can only be good to eat, pleasant to the palate and nutritious for the organism. Taste becomes at the same time an ethical and aesthetic parameter, a behavioural guideline and the reasonable application of the new scientific knowledge. Reason and taste, experiment and "tasting" are fused: the deer brains are the object of scientific observation and a delicious dish that slips from the experiment table to the pan, sizzling in the virgin lard of knowledge that certifies its tastiness and nutritious properties. The palate becomes a cognitive instrument no less sensitive and reliable than the experimental instruments with which nature is routinely tested.

Today however, public demonstrations such as the ones mentioned above in regards to bird flu seem to address the increasing insecurity and scepticism that characterise public opinion on these issues. By publicly eating chicken, politicians elude the sophisticated abstraction of data, the probabilistic risk estimates, and the often-differing estimates provided by experts. By risking their own safety — just as the great biologists used to try their vaccines out on themselves first — politicians

---

[28] Letter dated 29 September 1689 to Mr Jacopo Del Lapo, *Opere di Francesco Redi* (1809–1811, VI, p. 194).

aim to directly provide the reassurance that expert advice and policy measures can no longer guarantee.

Finally, these situations should be seen in the framework of a wider transformation of the nature of public visibility of political leaders. In particular, they can be interpreted based on the crisis of a founding tradition of modern politics, that of "the double body of the sovereign: one natural, physical and mortal; the other political, consecrated and immortal" (Boni, 2002, p. 24; cf. the classic Kantorowicz, 1957). A tradition that was capable of transfiguring sovereigns into the "miracle-worker kings" studied by Marc Bloch — who according to medieval popular belief were capable of healing Scrofula by the laying on of hands — which lead to the celebration of a double funeral rite, one for the physical body and another for the political one represented in portraits and sculptures (Ricci, 1998).

Even the newborn modern science played a role in updating and redefining the rituals linked to the sovereign's body and his miracle-working properties. In England at the time of the Restoration, around 1670, the ritual of the "royal touch" with which sovereigns cured people from scrofula was increasingly questioned. But thanks to the Royal Society, at the time also in need of legitimisation having been denounced by physicians, Charles II was able to associate himself with the public demonstration of a new liquid capable of healing wounds, and thus provided a new version, no less theatrical but more secularised and distant from the miracle-based forms of religious tradition (Werrett, 2000).

Today, in a context of pervasive media visibility that scrutinises and investigates every weakness of it, the body of the leader is increasingly "desacralized": "leaders without a political body, that are therefore, not called to identify themselves with a machine — of a party or a State — that guides their action and projects it beyond their life span. Leaders (...) with nothing else than their own body" (Calise, 2000, p. 112). Leaders therefore who have nothing left — for speaking to the public about emergencies such as the bird flu — than their own direct testimony, incarnated by biting into and chewing chicken in public.

Metaphorical and Epistemological Chickens

In the 1955 film *Rebel Without a Cause*, James Dean challenges other young men to the extremely dangerous "chicken game". The challenge consists in driving at high speed towards a cliff. The one who jumps out of the car first to avoid falling off the cliff is the "chicken" and loses. A variation of this game, the "chicken race", requires the two contestants to drive their cars towards each other. The "chicken" is the first one to jump out to avoid the impact.

This situation soon became part of the dilemmas studied in game theories, a branch of mathematical studies developed from the works of John von Neumann at the end of the 1920s. Just like the more famous prisoner's dilemma[29] the chicken game led to the analysis of possible competition and cooperation strategies between two subjects and the results that derived from them. For each participant in the chicken game, "the best result is to hold out to the end ("I compete") and for the other to veer out of the way ("he cooperates"). It is somewhat worse for me if we both get out of the way, because although I remain alive, the two of us are chickens" (Mero, 1998, p. 60). If neither gives up on the competitive choice, thus do not deviate or jump out of the car, the outcome is disastrous for both.

In this case therefore, chicken in science becomes a metaphor for human behaviour imprinted on caution (or cowardice) and on

---

[29] Introduced by Albert Tucker in 1951, the prisoner's dilemma focuses on the case of two criminal accomplices, imprisoned and interrogated separately. If one of the two confesses and not the other, the first will go free and the second will be condemned to a harsh sentence. If both confess, they will both get a more moderate sentence. If neither confesses, they will both receive a mild sentence (cf. Mero, 1998).

cooperative tendencies, in opposition to impudent (or insane) courage and competitive tendencies.[30]

Unlike the prisoner's dilemma, whereby it is best to compete notwithstanding what the adversary does, in the chicken game it is best to cooperate (therefore deviate the car) if the adversary competes and vice-versa. "The player who cannot bear the risk of the worst outcome is a certain loser in such games" (Ibid., p. 61). This is particularly obvious if one transfers the "chicken game" to conflicts between great powers. At the beginning of the Second World War, Hitler understood that Prime Minister Chamberlain was not ready to enter into war and he therefore prevailed in a series of "chicken situations". On the other hand these situations tend to lead to cooperation when each of the parties involved succeeds in convincing the other that they have no intention of cooperating, and that they are ready to compete until the end. This is what happened in 1962 during the so-called "Cuban missile crisis" which lead the two superpowers to the brink of World War Three. With the help of consultants who knew game theories well, Kennedy was clear in conveying to Khrushchev that the United States would not have backed down even faced with the threat of a nuclear conflict against the Soviet Union.

Finally, sociologists of science Harry Collins and Steven Yearley (1992) use the expression "epistemological chicken" to describe a similar escalation in the field of science studies, a sort of "competition to be the most radical" that is at times one of the most widespread temptations in these types of studies, a race, which, according to both sociologists has particularly negative consequences.

---

[30] It is very unlikely that this characterisation, like the more widespread one of the chicken being an animal with scarce cognitive capacities ("to be a bird brain"), gives justice to the actual behaviours and physiology of these animals (cf. Vallortigara, 2005). In terms of being domesticated or family pets, chickens are mentioned in numerous traditional expressions, such as "to know one's chickens", "to be cocky (a show off)", "to run around like a headless chicken (someone who is indecisive)" (cf. Lapucci, 1979).

Other Birds:
Redi's peacock, Benjamin Franklin and the economists' turkeys
(very similar to Bertrand Russell's chicken)

In addition to chicken, there are other feathered animals that fly between the dining table and the world of scientific research.

According to Piero Camporesi, the significant change in taste of the privileged classes between the XVIIth and XVIIIth centuries lead to peacock meat — once considered "delicious" — being abandoned in favour of turkey meat. The above-mentioned Francesco Redi received one from his friend Lorenzo Magalotti, another man of letters and science, with a recipe in verse:

> Your taste is so accomplished and fine
> That I am sure you will know how to cook it
> You are a doctor not a bird brain
> And for sure will boil it or eat it in sauce[31]

(One should note the contrast between "doctor" and "bird", the latter clearly already considered a symbol of poor intellectual capacity.)

In his *Science in the Kitchen*, more than two centuries later, Pellegrino Artusi does not mention recipes containing peacock meat, but dedicates an interesting *excursus* to this bird, (*Pavo Cristatus*, recipe 550), a source of "excellent meat for young people", and defines it "the most splendid of the gallinaceous birds, for its magnificent display of colors"; he tells the story of how Alexander the Great, who brought it into Greece from Southeast Asia, had been so struck by the beauty

---

[31] Cf. Camporesi, introduction to Artusi (2001, p. LXXV; original ed. 1891). The quote is from L. Magalotti, *To Mr Francesco Redi*, in *La donna immaginaria* (*The Imaginary Woman*), 1762, cit. in Camporesi (1990, new ed. 1998, p. 52).

of these animals "he forbade them to be killed, under pain of severe punishment" (Artusi, 1891, trad. Eng. 2003, p. 383).

It is thought that this is how turkey made its way onto menus and into scientists' consideration. Of course it holds a special place in the history and cooking traditions of the United States: to the point where Benjamin Franklin stated that it would have been better to choose a turkey instead of an eagle as the symbol of the United States of America (Dubourcq, 2000, new ed. 2004, p. 62).

Turkey is often referred to in the field of economics and in broader reflections on probabilities and forecast models.

The "undercover economist" Tim Harford mentions "turkeys voting for Thanksgiving" to explain a series of price-targeting strategies that encourage clients themselves to indicate the highest price they are willing to pay for an item. For example according to Harford, large international chains offer coffee in slightly different doses and characteristics (for example giant cappuccinos and moka coffee), of which the costs of preparation are probably almost identical, but which enable them to "flush out" the clients who are willing to pay more. This way, the consumer "gives himself away" against his own interests, exactly like a turkey were it to vote for Thanksgiving (Harford, 2006, 2012, p. 42).

The mathematician and philosopher Nassim Nicholas Taleb, author of a book with another bird in the title (*The Black Swan*), sees in the turkey and its destiny at Thanksgiving (or at Christmas) a metaphor of our incapacity to foresee dramatic events (such as serious accidents or a stock market crash) based on former experiences and expectations of regularity. A turkey that is fed every day tends to develop a sense of security and trust in human beings, until its abrupt ending on the Wednesday preceding Thanksgiving. Paradoxically, Taleb explains, the turkey's "sense of safety" "reached its maximum when his risk was at its highest!" (Taleb, 2007).

As indicated by Taleb, this is an "Americanised" version of the theme of induction previously analysed by Bertrand Russell, based on — surprise surprise — the example of a chicken.

> Experience has shown us that, hitherto, the frequent repetition of
> some uniform succession (...) has been a cause of our expecting

the same succession or coexistence on the next occasion. Food that has a certain appearance generally has a certain taste, and it is a severe shock to our expectations when the familiar appearance is found to be associated with an unusual taste. (...) And this kind of association is not confined to men; in animals also it is very strong (...). The man who has fed the chicken every day throughout its life at last wrings its neck instead, showing that more refined views as to the uniformity of nature would have been useful to the chicken. (Russell, 1957, 1998, p. 34)

DRINKS ON THE SIDE

Beer, Wine, Coffee, Tea, Chocolate and… as Many Controversies as You Like

What a problem breakfast was! They gave up white coffee, because of its terrible reputation, and then chocolate, for it is "a mass of indigestible substances". So there remained tea. But "nervous people should shun it completely".

GUSTAVE FLAUBERT
Bouvard and Pecuchet

Beer, Reason and Work According to Benjamin Franklin

In 1725 Benjamin Franklin is not yet the great scientist and politician that he was later to become, but a young curious man and avid reader who dreams of opening a printing company in Philadelphia. Trusting the promise made to him by the governor, he travels to London to acquire the necessary material for his venture. Young Benjamin is attracted to the contacts and intellectual exchanges that the English capital offers: he meets Bernard de Mandeville, the author of *The Fable of the Bees* and hopes to be introduced to Newton, of whom he is a great admirer.

But things take a negative turn: the trust he has placed in the governor turns out to be misplaced and Franklin finds himself stuck in London without the necessary funds for his return trip. He therefore adapts and starts working for a printing company, where he is shocked by the diet and especially the drinking habits of his English colleagues.

> I drank only water; the other workmen, near fifty in number, were great guzzlers of beer. On occasion, I carried up and down stairs a large form of types in each hand, when others carried but one in both hands. They wondered to see, from this and several instances, that the *Water-American*, as they called me, was *stronger* than themselves, who drank *strong* beer! We had an alehouse boy who attended always in the house to supply the workmen. My companion at the press drank every day a pint before breakfast, a pint at breakfast with his bread and cheese, a pint between breakfast and dinner, a pint at dinner, a pint in the afternoon about six o'clock, and another when he had done his day's work. I thought it a detestable custom; but it was necessary, he suppos'd, to drink *strong* beer, that he might be strong to

labor. I endeavored to convince him that the bodily strength afforded by beer could only be in proportion to the grain or flour of the barley dissolved in the water of which it was made; that there was more flour in a pennyworth of bread; and therefore, if he would eat that with a pint of water, it would give him more strength than a quart of beer. (Franklin, 1868, pp. 146–147, italics original)

Can you imagine young Franklin trying to explain to the inebriated worker, that the "physical strength" that the organism derives from beer is necessarily "in proportion to the quantity of barley"! Yet what we witness here is the first stage of that same mechanism that was used during the television show dedicated to mayonnaise (cf. chap. *Starter*), during the demystification of naïve cooking practices in the case of the "Enlighteners' chicken" (cf. chap. *Main Course*) and in numerous other contexts where science got in contact with food practices and their common representations. The worker, intoxicated with beer is victim of a pre-scientific *misconception* whereby he confuses an apparent and superficial similarity (strong beer gives me more strength to work) with a substantial characteristic. An enlightened and rational spirit such as Franklin cannot accept that such misconceptions continue. He therefore "breaks down" the ingenious worker's theory with instruments that are typical of scientific investigation: the flour and the water contained in the beer are what provide energy to the organism, therefore it would be much more rational to eat bread and drink water!

How does the worker react to Franklin's explanation? He takes no notice, as one could have imagined. He continues drinking and "paying every Saturday night a score of four or five shillings a week for this cursed beverage; an expense from which I was wholly exempt. Thus do these poor devils continue all their lives in a state of voluntary wretchedness and poverty".

Until suddenly, something happens. The colleagues demand that Franklin also contribute five shillings for buying beer. Franklin considers this an injustice but after a series of harassments from his work colleagues — they mix up the pages he has already typeset, break the press characters — he resigns himself to pay his due for the beer.

I was now on a fair footing with them, and soon acquir'd considerable influence [...] From my example, a great part of them left their muddling breakfast of beer, and bread, and cheese, finding they could with me be suppli'd from a neighboring house with a large porringer of hot water-gruel, sprinkled with pepper, crumbl'd with bread, and a bit of butter in it, for the price of a pint of beer, viz., three half-pence. This was a more comfortable as well as cheaper breakfast, and kept their heads clearer. Those who continued sotting with beer all day, were often, by not paying, out of credit at the alehouse, and us'd to make interest with me to get beer; *their light*, as they phrased it, *being out*. I watch'd the pay-table on Saturday night, and collected what I stood engag'd for them, having to pay sometimes near thirty shillings a week on their account. (Ibid., pp. 148–49, italics original)

The reasons of science therefore come to coincide with those of morality and common sense, including the ethics of saving. The workers who continue to guzzle beer are "poor devils" who end up fat and with an empty wallet by the end of the week. Franklin describes them with a wonderful metaphor, at once technological, religious and mundane: "their light had gone out", the workers have lost their senses, the dignity of a thrifty and sensible worker that is typical of the protestant culture. Gruel, toasted bread and butter are more nutritious than beer, keep the brain lucid and even help save money! In the end, Franklin's pragmatism eventually succeeds in getting reason to prevail and enables him to earn a little money by loaning it with interest.

These themes are recurrent on various occasions and in various forms during our journey through the interactions between science and cooking. They are often reinforced by another theme, which is that of "good taste" that emerges as a broader culinary and cultural category. The different styles and culinary diets that are recommended and listed in various manuals are defined and justified by this initial scientific knowledge, moral and culinary expectations and "good taste" — which in part contributes to summarising them all.

One of the leading Italian Enlightenment intellectuals, Pietro Verri, describes the table of a "man of good taste who seeks the truth" as one from which "*hard drugs*, salted foods, truffles and similar venoms of human nature" are "entirely forbidden"; where flavours are "delicious but *not strong*; any food that impacts strongly on the palate makes the palate itself more or less lazy, and deprives it from an infinite number of more delicate pleasures"; most strong tasting foods affect "the lining of the ventricle and of the intestines, and this causes an infinite number of ailments that negatively compensate the pleasure of the sensation". Even wines should have "*strong taste and little strength*" to aid digestion. "No *strong smelling* food should be allowed on our table, and any herb that emits a bad smell when entering in contact with water should be banished from the table therefore persimmons, and any type of cabbage should be excluded. In such a way our dinner, which we end with an excellent cup of coffee, will leave us satisfied and sated, and not oppressed by heavy nutrition".[32]

Contrary to the London printers cited by Franklin, the adjective "strong" always carries a negative connotation in this case. Flavours, spices and strong wines are banned; digestion should be easy for the brain to stay clear. For an intellectual such a Chamfort (1741–1794), a light meal goes hand in hand with politeness and science in the kitchen.

> When one sees on the table light dishes, wholesome and well prepared, one is happy indeed that cooking has become a science. But when one sees gravies, rich bouillons, truffled pâtés, one curses the cooks and their morbid art: it is all in the application. (Chamfort, 1795, Eng. trans. 2014, p. 3)

Measure and good taste also mark the distance from the excesses of the court kitchens of the courts during the previous centuries. The favourite drink is no longer wine, nor beer, but coffee. The latter "awakens the mind (…) and it is established that it infuses the blood with a volatile salt that accelerates its flow, as well as clearing, thinning and in a way

---

[32] Verri (1766, p. 201, italics mine). Camporesi also mentions this extract (1990, new ed. 1998). On good taste also see Montanari (2009b). In the entry "Goût » of the *Encyclopédie*, Voltaire makes an explicit parallel between culinary taste and the capacity for aesthetic judgment, and between *gourmet* on one hand and *homme de goût, connaisseur,* on the other. (*Encyclopédie*, p. 161). See also Sermain (1999) and Mangione (2003).

reviving it (…); it is particularly useful to people who practice little movement and who study sciences" (Verri, 1766, pp. 7–8).

Almost two centuries earlier however, the virtues of beer had been much appreciated by the scholar Francis Bacon who had no need to stimulate his intellect, in fact if anything he needed much the opposite: to placate it.

> His Lordship would often drinke a good draught of strong beer (March beer) to-bedwards, to lay his working fancy asleep: which otherwise would keepe him from sleeping great part of the night. (Aubrey 1950, p. 129)

Beer, Bacteria and Beetroot:
Louis Pasteur's fermenting science

Everything started with a beetroot. In the summer of 1856, Louis Pasteur is reaping the first recognitions of his long and successful career. Nominated a couple of years earlier as Headmaster of the newborn Faculty of Sciences in Lille, in his acceptance speech he has immediately underlined the practical importance that scientific studies could have for the local industry in which, among other things, numerous bars and distilleries were active.

Among Pasteur's students there is a certain Bigo, the son of an important beetroot-based alcohol distillery owner. His desperate father calls upon Pasteur for help: something is wrong; the alcohol production is slowing down. Pasteur goes to the distillery various times and takes some samples from the vats that are still active in the alcohol production and others from the now idle ones. The young Bigo would later write in a letter that under the microscope "Pasteur had noticed (...) that the globules were round when the fermentation was healthy, that they lengthened when alteration began, and were quite long when fermentation became lactic". "When studying the causes of these failures" continued Vallery-Radot, Pasteur's biographer, "Pasteur had wondered whether he was not confronted with a general fact, common to all fermentations. Pasteur was on the road to a discovery the consequences of which were to revolutionize chemistry" (cit. in Geison, 1995, p. 93).

This was probably another case of an anecdote being stressed by biographers and (some) historians[33] but there is no doubt that his communication in 1857 entitled *Mémoire sur la fermentation appelée*

---

[33] Among these, beside Vallery-Radot, Dubos and the great success of De Kruif (1926), but also J.D. Bernal (1953).

lactique ("*A memoir on the Fermentation called Lactic*") marks the first firm step in what was then to lead to his research "from crystals to life". In it, he fervently lays out the hypothesis that fermentation might not be a purely chemical process, but that it is, among other things, the result of living microorganism activity.

This is a victorious landmark in his long challenge against his afore-mentioned German rival Justus von Liebig (cf. chap. *Starter*), who in the past, with other chemists such as Friedrich Wohler had harshly and sarcastically opposed similar hypotheses previously formulated by French engineer Charles Cagniard de la Tour and chemist Theodor Schwann. De la Tour had taken beer samples from a few distilleries and noted some "buds" similar to germinating seeds in the small yeast globules (De Kruif, 1926, 1954, p. 58). But the scepticism of Wohler and Liebig got to the point of publishing, as a footnote to an article in which they speculated on the role of microorganisms in fermenta-tion, a satirical note on the observation under a microscope of "wine animals" in the "shape of a Beindorf distilling flask", that "eat sugar, eliminate alcohol from the intestinal tract, and carbon dioxide from the urinary organ. The urinary bladder in its filled state has the shape of a champagne bottle..." (1839 cit. in McGee, 1984, p. 434). For Liebig, this hypothesis was comparable to "the opinion of a child who would explain the rapidity of the Rhine current by attributing it to the violent movement of the many millwheels at Maintz" (Dubos, 1960, p. 82).

For Pasteur this is the beginning of years of frenetic activity divided between research, teaching, intense relationships with farmers and local businesses and visits with his students to the distilleries of Valenciennes.

We have mentioned how, in line with a well-established public image, his biographers emphasised his indifference for food during crucial moments of his research — food was something of a mere distraction to be reduced to a minimum or to be avoided altogether (cf. chap. *Main Course*).

The competition between scientific and cooking practices is also exalted by the quasi "domestic" context of these experiments, that often share spaces with the traditional places of food and drink preparation.

In 1858, during his traditional holiday in Arbois, Pasteur makes assiduous use of the well-supplied cellars of his childhood friends for his microscope observations on deteriorated wine, and notes strong resemblances with what he had observed in lactic yeast.

> He set up his laboratory in what had been *an old café* and instead *of gas burners he had to be satisfied with an open charcoal brazier that the enthusiastic Duclaux kept glowing with a pair of bellows;* from time to time Duclaux would scamper across to the town pump for water; their clumsy apparatus was made by the [in expert hands of the] village carpenter and tinsmith. (De Kruif, ibid., p. 92, italics mine)

It is like witnessing the scene of daily practical life, a representation from a nativity scene: the abandoned café, the assistant running back and forth with buckets of water then using bellows to keep the fire going. This is still a "homemade science", in places and with instruments that are improvised, and often indistinguishable from those used for other practices — starting with the kitchen.

Other than yeast, Pasteur regularly observed traces of microorganisms in the samples of deteriorated wine that were absent in the non-deteriorated ones. "So skillful did he become in the detection of these various germs that he soon was able to predict the particular flavor of a wine from an examination of the sediment" (Dubos, 1960, p. 67).

> Then he called the winegrowers and the merchants of the region together and proceeded to show them magic. "Bring me a half dozen bottles of wine that has gone bad with different sicknesses", he asked them. "Do not tell me what is wrong with them, and I'll tell you what ails them without tasting them". The winegrowers didn't believe him; among each other they snickered at him as they went to fetch the bottles of sick wine: they laughed at the fantastic machinery in the old café, they took Pasteur for some kind of earnest lunatic. They planned to fool him and brought him bottles of perfectly good wine among the sick ones. (…) With a slender glass tube (Pasteur) sucked a drop of wine out of a bottle and put it between two little slips of glass before his microscope. The wine raisers nudged each other and winked French winks of humorous common sense, while Pasteur sat hunched over his microscope, and they became more merry as

minutes passed (…). Suddenly he looked at them and said: "There is nothing the matter with this wine — give it to the taster — let him see if I'm right". The taster did his tasting, then puckered up his purple nose and admitted that Pasteur was correct; and so it went through a long row of bottles — when Pasteur looked up from his microscope and prophesied: "Bitter wine" — it turned out to be bitter; and when he foretold that the next sample was ropy, the taster acknowledged that ropy was right! The wine raisers mumbled their thanks and lifted their hats to him as they left. "We don't get the way he does this — but he is a very clever man, very, very clever" they muttered. That is much for a peasant Frenchman to admit…. (De Kruif, ibid., pp. 92–93)

Like in the case of mayonnaise (cf. chap. *Starter*), science proves it is capable, with its own methods, of achieving more and better results compared to established practices and experiences such as those of the taster and the vine growers who have to rely on subjective sensations (smell and taste) in order to recognise an acidic wine. Thanks to this demonstration, Pasteur wins over the scepticism and the initial scorn of the winegrowers and science thus triumphantly enters the cellars, legitimising its role in this field. But one of the most notable aspects of this situation — and one that would have heavy consequences that continue to this day to weigh on the public image of science — is that it does not actually achieve this triumph *scientifically*. Pasteur gives no explanation whatsoever to the vine growers, on his method or the theories on which it is based. The vine growers go back home incredulous and full of admiration without having understood anything, just like the ladies who witnessed the mayonnaise debacle (cf. chap. *Starter*). Science is offered here as a *black box*, with results that are accepted and appreciated without being understood. Just as the biographer underlines, the demonstration is very similar to a "magic trick", and such is the role of science in this circumstance, as described in anthropological terms by Malinowski as taking on a "directed towards the attainment of practical aims. (…) intimately associated with human instincts, needs, and pursuits" (Malinowksi, 1948, p. 66).[34]

---

[34] For a more general analysis of this topic, see Bucchi (2010a).

During the following years, Pasteur extends his observations to include beer. It is a practical necessity, not just financial but also political this time, that contributes to him focussing his research on the topic. After the war with Prussia, it is crucial to revamp the role and the glory of France, and what more significant challenge than that of improving the quality of beer, a symbol of the rival nation? For a while, Pasteur transforms his laboratory into a "model distillery", and in 1876 he writes the treaty *Etudes sur la bière, ses maladies, les causes qui les provoquent, procédé pour la rendre inaltérable, avec une théorie nouvelle de la fermentation*. The conclusion is once again in line with his hypothesis, that the alterations "in the wort, and in the beer itself, are due to the presence of microscopic organisms of a nature totally different from those belonging to the yeast proper" (Dubos, 1960, pp. 67–68).

To tackle this problem, based on his experiences with wine, Pasteur perfects the thermal treatment that later will be known as pasteurisation, during which he manages to partially sterilise drinks without altering their taste or smell. One of the most interesting aspects, noted by some of his biographers, is that in other culinary fields, procedures of "controlled heating" and food preserving in bottles and tins had been introduced since 1810.

> [...] so that it may seem that Pasteur was demonstrating the obvious. But there is always a lag in many scientists' apprehension of the implications of the practical man's work. One could say facetiously that Pasteur had to demonstrate to scientists what the housewife already knew. All the examples mentioned show how much had been discovered by trial and error long before the days of experimental science. What Pasteur brough to the problem was the concept that most food spoilage is caused by microorganisms of various kinds, and he thus provided a new understanding that gave increased significance to the ancient empirical techniques. (Dubos, 1960, pp. 71–72)

In this case too therefore, science comes to systematise and ground already widespread knowledge and practices from a theoretical and experimental point of view, by getting "scientists and housewives" to agree. Incidentally, a very similar dynamic would characterise Pasteur's studies on vaccinations: the concept of immunisation (as in to provide protection from an illness through a lighter form of the same illness) and the practice linked to inoculation had already been familiar to popular culture for a while (Bercé, 1984; Darmon, 1986; Bucchi, 1998a). According to a reconstruction accepted by many historians, even Jenner — the first doctor to give scientific dignity to the vaccination by introducing it as a preventative method against measles — was encouraged to undertake his research after a series of conversations with the female farmers in the Scottish countryside. Because they worked in close contact with cows, the farmers had noticed that the development of a light form of infection (the measles vaccine) contracted by the animals later rendered them immune to human measles (cf. Dubos, cit.).

A Sip of Coffee, and One Sip of Controversy

On the 16th January 1680, patrons at Garraways Coffee Shop in London interrupt their conversations due to a rather unusual scene taking place. Whilst their coffees are still steaming, two men briskly fiddle with a set of strange utensils. One of them is Robert Hooke, curator of experiments at the Royal Society, one of the first scientific academies. Formerly an assistant to Robert Boyle, Hooke is now in charge of coordinating and preparing the weekly experiments that members of the society discuss every week.

During the previous months, Hooke, whose interests range from physics to natural history, has been corresponding with Newton on the subject of the fall of bodies. Hooke decides to undertake an experiment indoors, to avoid the influence of drafts. As a regular customer of coffee shops, Hooke decides to conduct the experiment at Garraways, the main hall of which is 27 feet high, or approximately eight meters. With the help of his assistant Harry Hunt, Hooke attaches a plumb line from the ceiling, then drops one of the bullets and measures its south-eastern oscillation, thus demonstrating the "daily movement of the Earth", as he notes in his diary on January 22nd.

Hooke reports his results to the Royal Society, where "it was desired that this experiment might be made before a number of the Society, who might be witnesses of it, before the next meeting" (Hooke, cit. in Inwood, 2004). The following Monday, they return to Garraways with Sir William Petty to repeat the experiment.

On 28th April 1686, just over six years later, Newton's masterpiece *Philosophiae Naturalis Principia Mathematica,* in which he laid out his theories on the laws of dynamics and gravitational motion is presented to the Royal Society. At the end of the meeting, Hooke, Halley and some other colleagues go to the Grecian Coffee Shop, where Hooke claims paternity of the discovery, maintaining that he has given Newton the idea first, in particular of the "importance of Kepler's proportions and the Inverse Square Law" (Chapman, p. 209). "Being adjourned to the coffee-house," Halley wrote to Newton on 22 May, "Mr Hooke did there endeavour to gain belief, that he had such thing by him, and that he gave you the first hint of this invention." But Hooke's efforts were useless: the coffeehouse had given its verdict, which still stands today (Standage, 2005, p. 119).

The quarrel between Hooke and Newton is a classic example of a dispute over the right of priority, a topic which is often and widely studied by historians and sociologists of science.

Robert K. Merton, the founder of the sociology of science, regarded these disputes as bearing particular significance when seeking to identify the norms that characterise science as an institution. Merton documented numerous disputes concerning priority involving great scientists, from Galileo to the above-mentioned Newton and Hooke, from Cavendish to Faraday, from Laplace to Legendre, Gauss and Cauchy. Merton quickly dismisses the idea that the quarrels arise from generic human egocentricity, or from science's tendency to attract particularly querulous characters. According to Merton, the frequency of such disputes is a consequence of the very priorities of science as an institution — i.e. the emphasis on originality being key to the development of knowledge — and to the mechanisms that provide an incentive for the single scientist to contribute to these priorities.

"(...) The *great frequency* of *struggles* over *priority* does not result merely from these traits of individual scientists but from the institution of science, which defines originality as a supreme value and thereby makes recognition of one's originality a major concern" (Merton, 1973: 294). It is not a coincidence that from the outset, scientific academies such as the Royal Society had established the tradition of "having sealed and dated manuscripts deposited with them in order to protect both priority and idea" (Merton, 1973: 364).

But how is it possible for two or more scientists to reach the same discovery or very similar conclusions more or less simultaneously? To answer this question, Merton refers to Francis Bacon, having found an "implicit social theory of discovery" in his writings, based on four elements. The first three are the "incremental accumulation of knowledge, the sustained social interaction between men of science and the methodical use of procedures of enquiry". The fourth element describes all innovations, including scientific ones, as "births of the times", and Time as the "greatest innovator" (Bacon, cit. in Merton, 1973, p. 349).

As for other themes, Merton does not hesitate to accumulate data and examples by documenting the extensive engagement of the greatest scientists in cases of "multiple discoveries", i.e. discoveries made simultaneously by two, and often three or more scientists: Galileo, Newton, Faraday, Maxwell, Gauss, Laplace, Lavoisier. After analysing 400 scientific communications by Lord Kelvin, Merton reckons they contain at least 32 multiple discoveries. Merton does not consider this phenomenon to be pathological, or abnormal, but rather he finds it intrinsic to the very organisational practices of science, and as such well known to scientists themselves. Amongst the many cases referred to, there is the one involving Gauss. When in 1795, at the age of 18, Gauss elaborates the method of Least Squares, "to him the method seems to flow so directly from antecedent work, that he is persuaded others must already have hit upon it" (Merton, 1973, p. 363). Merton's conclusion is only slightly paradoxical: "it is the singletons — made only once in the history of science — that are the residual cases, requiring special explanation" (Merton 1973, p. 356).

This conclusion does not however preclude the role of creativity and individual talent. "By conceiving genius sociologically as one who in his own person represents the equivalent of a number and variety of often lesser talents, the theory maintains that the genius plays a distinctive role in advancing science, often accelerating its rate of development and sometimes, by the excess of authority attributed to him, slowing further development" (Merton 1973, p. 370). This partly explains the strange quarrel between Newton and Hooke, and the sour taste that it left in Hooke, which could not even be sweetened by his beloved coffee.

There are, however, situations which relate more directly to the substance of scientific content, and not only to the paternity of a discovery or a result attributed to one specific scientist.

A scientific controversy is a situation in which the knowledge in a certain domain or phenomenon has not yet been stabilised.[35] Scholars who study science and its relationship with society believe that controversies are particularly interesting and that they offer relevant opportunities for analysis. In some cases for example, controversies have led to the understanding of explicit or implicit mechanisms in attributing credibility or faith to certain subjects or institutions and thus to their statements and results. To cite a well-known scientist, Pasteur took part in a great number of fierce controversies, including the one regarding spontaneous generation which set him in opposition to his colleague Pouchet. Pasteur's hypothesis eventually prevailed thanks to his famous high-altitude experiment. A commission, put together by the Académie des Sciences, issued a verdict which was favourable to Pasteur, and rejected Pouchet's theory on spontaneous generation.

Controversies amongst scientists are often linked to debates within society that are just as fiery.[36] Disagreements amongst experts acquire a higher degree of visibility when there is a simultaneous social and political debate on the potential implications of a techno-scientific issue, e.g. contemporary discussions on GMOs or nuclear power. There are many reasons to believe that this kind of controversy is more frequent — or at least more visible — in the current framework of relations between science, politics, media and citizens. However it would be a mistake to believe that this is a radically new situation. Society's interest and involvement in a certain number of scientific debates is a recurring fact, although in time it has taken on different forms, particularly when the issues at hand concern aspects bearing practical interest for daily life such as health or food.

We have already referred to Liebig, who succeeded in resisting the scepticism of doctors and chemists regarding the nutritional properties of his meat extracts, thanks to the support of housewives (see chap. *Starter*).

But no less interesting is the controversy on coffee, which accompanied the rapid diffusion of the drink in Europe during the second half

---

[35] For a general introduction on the subject and the analysis of scientific controversies, cf. Callon (1981), Bucchi and Lorenzet (2008), Bucchi (2010b), Lorenzet (2011).

[36] Cf. Martin and Richards (1995), Bucchi (2006).

of the XVIIth century. In 1674, a group of women in London signed the *Women's Petition Against Coffee Representing To Publick Consideration the Grand Inconveniences Accruing To Their Sex From The Excessive use Of That Drying Enfeebling Liquor.* According to the signatories, the "drying and enfeebling liquor" compromised the sexual capacities of their husbands, rendering them "as unfruitful as the deserts, from where that unhappy berry is said to be brought" and leading them to "trifle away their time, scald their Chops and spend their money, all for a little base, black, thick, nasty, bitter, stinking, nauseous puddlewater" (Standage, cit., p. 106, M.E. Snodgrass, 2004, p. 236).

The petition was answered in an anonymous document by a group of men who argued that coffee had the capacity to promote "vigorous erections and full ejaculations", supporting their claims by referring to the works of physiologist William Harvey, who had praised the capacity of coffee to lubricate blood circulation (Snodgrass, 2004, p. 236). Testimonies from those times confirm that the discoverer of blood circulation William Harvey "was wont to drink *coffee*; which he and his brother *Eliab* did, before *coffee*-houses were in fashion in London" (Aubrey, 1950, p. 139).

Another more ambivalent point of view came from a famous Oxford physician, Thomas Willis (1621–1673), who categorises coffee as an "anti-hypnotic" in his chapter on opiates. According to Willis, coffee can be "profitable" in some cases, whilst in others it is "hurtful and unwholesome, for we see that great coffee drinkers become lean and are very often subject to be paralytic, and grow impotent for generations". However, Willis admits to prescribing it often. "For indeed in very many cephalic diseases and infirmities, (such as) headaches, giddiness, lethargy, catarrh and the likes (…), coffee has often a very good effect, for being daily drunk, it wonderfully clears and enlightens each part of the Soul" (Willis, 1685, *The London Practice of physic*, pp. 68–69).[37] In Marseille, when the first French coffee shop opened, the shop managers who feared competition from the new drink, requested a note from local

_____

[37] On this topic see also Bresadola and Cardinali (2009). Other interesting controversies like the one regarding sugar are analysed in Fischler (1992).

doctors who maintained that coffee "induces palsies, impotence and leanness" (Standage, cit., p. 107).

The Swedish naturalist Carl Linnaeus, who classified the coffee plant in his system as *Coffea Arabica*, was a great coffee drinker and unsuccessfully tried to cultivate the plant in Sweden. During his time as doctor of the Admiralty, he drank a cup of unsweetened black coffee to cope with some of his patients' asphyxiating halitosis. In one of the dissertations written for his students, *Potus coffeae* in 1761, Linnaeus discusses the effects of coffee but also describes some of the risks linked to its excessive consumption which:

> "promotes awakeness and destroys the appetite… it is said to prevent flatulence and help digestion after a meal; but there is no doubt that coffee is not sufficiently toasted, it *induces* flatulence and belly rumblings…Most confirmed and regular coffee drinkers have trembling hands and heads. I remember three distinguished person (now dead) who from over-indulgence in coffee could scarcely get their shaking hands to their mouths. However, in time and by abstinence, they happily were cured. One of these, who could not deny himself the taste, used to hold the coffee in this mouth and then spit it out instead of swallowing it". (Linnaeus, 1761, cit. in Blunt, 1971, p. 157)

Linnaeus also claims that coffee has "anti-aphrodisiac" properties, as mentioned by the London women. On this topic he quotes the story told by a German traveller according to whom the wife of a sultan, horrified at having to castrate a riotous horse, proposed to solve the problem with large quantities of coffee, the benefits of which she had already seen on her husband. Similar stories contribute to giving coffee the derisive nickname of the "eunuch's drink" (Blunt, cit. 157–158).

In general, XVIIth and XVIIIth century scholars attribute to coffee the property of drying up bodily fluids and in particular the so-called "phlegm" — according to the British physician Benjamin Moseley[38] (1742–1819) — and thus it was described as particularly indicated for phlegmatic temperaments and for "people with large, corpulent

---

[38] Moseley (1785).

physiques, who are very prone to catarrh", as also illustrated in the *Encyclopédie* (Moseley, cit. in Schivelbusch, 1980; Eng. trans. 1992).

One of the most interesting aspects however, is that science reaches its conclusions on coffee through analogy rather than through actual scientific method — coffee "dries up" because it is a hot drink and because of the roasting process — and particularly through social and cultural affinity with the new drink and with the places in which it was drunk.

Coffee quickly becomes associated with leanness, sobriety and concise reasoning; it is considered useful for the new intellectual activities such as science: "particularly useful to the people who study sciences" according to the above-mentioned Verri. According to the Swiss doctor Samuel Auguste André David Tissot[39] (1728–1797), when drunk in moderation, coffee "causes a cheerfulness of mind and [increases its penetration; for which reason the learned are so fond of it" (Tissot, 1768, Eng. trans. 1768, p. 145). Hence, coffee is opposed to beer and to phlegmatic temperaments, although some of its properties are wrongly emphasised; for example the misguided conception of coffee rendering one sober after excessive consumption of alcohol. Coffee is seen as having "great sobering properties"; it incarnates a new synthesis of "dry" and pragmatic rationality and sobriety of customs. It also symbolised the public character of the new knowledge which was growing within a new "caffeine-fueled" network (Standage, cit., p. 125).

Some coffee shops, in London particularly, soon became habitual meeting and discussion places for several of the main actors of the scientific revolution. Various members of the Royal Society such as the mathematician and physicist Isaac Newton, the astronomer Edmund Halley and the physician and naturalist Hans Sloane were patrons of the Grecian. In his diary, the above-mentioned Robert Hooke, a great *habitué* of these coffee shops, documents how many discussions with colleagues — and even experiments — took place in front of a cup of coffee.

---

[39] Tissot (1768, p. 145).

Between the end of the XVIIth and the beginning of the XVIIIth century, coffee shops became one of the places in which discoveries by the natural philosophers spread to a wider public. In 1698 John Harris gave a number of lecture on mathematics at the Marine café; initially free of charge, they soon attracted a paying public. The demonstrations undertaken by James Hodgson at the Marine were even more spectacular, and were publicised by emphasising the use of instruments that had been "rarely seen outside the Royal Society" (Jacob and Stewart, 2004, p. 76). Hodgson illustrated discoveries by scholars such as Boyle and Newton with the help of a vast selection of air pumps, microscopes, telescopes and prisms built by Francis Hauksbee, with whom he formed a very successful partnership.

During the convivial ceremony celebrating the prize that is most significantly associated with the public image of science — the Nobel prize banquet in Stockholm — since 1927 the brief thank you speeches made by Nobel laureates in the sciences and other fields take place after dessert, when coffee is served (Soderlind, 2005, Eng trans., 2010; see also Bucchi, 2017).

Tea vs. Coffee,
Or why kings should not be involved in clinical tests

The emerging of science as an institution between the XVIIth and XVIIIth centuries sees a great opportunity in the application of its developing analytic techniques to the new "exotic" food and beverages that were quickly spreading through Europe following the geographic discoveries and the colonial and commercial expansion processes. The emerging advice and their relevance to many daily activities — from food preparation to commerce — contributed to legitimise and substantiate the increasingly significant social role of men of science.

Scholars of the time offered for example their views on tea. Although he admitted the beverage had certain benefits, Simon Pauli, the King of Denmark's physician, warned of its poisonous and potentially mortal nature in an essay in 1653, and based his argument on geopolitics: he attributed the danger to the fact that it came from and was imported from China. Dutch physicians Nicolas Dirx and Cornelius Bontekoe, on the other hand, gave opposite and enthusiastic opinions. The latter wrote an essay on tea in 1678, which some suspected had been financed by the East India Company, in which he recommended "that every man, every woman drink it every day, if possible every hour" (Schivellbush, *Tastes of Paradise* (Pantheon Books, 1992, p. 80).

However, it was not a scientist, but a sovereign who first applied a curious "experimental" approach to the dilemmas linked to the new drinks. Gustav III of Sweden (1746–1792) was convinced that coffee was bad for the health and resolutely wanted to try out his "theory". In what is, ironically, considered to be one of the first clinical trials in history, King Gustav pardoned a death row prisoner on the condition that he drink coffee every day. In order to have some sort of "control" he pardoned a second man, and commanded that he drink tea. Two physicians were in charge of monitoring the health of the two human guinea pigs. But the two physicians actually died before the prisoners, and in 1792 the King was killed. The tea drinking man died at the age of eighty-three, before the coffee drinker, although it is not known how much longer the second man lived on. The result however was not taken into consideration, since coffee was banned in Sweden twice, in 1794 and in 1822.

Commenting on this peculiar story, contemporary epidemiologists find it difficult to reach a decisive conclusion from such an "experiment", given the restricted sample and the numerous external factors that could have potentially affected the experiment, except for one: a sovereign should not meddle with clinical tests.

Today, researchers continue to question — albeit based on studies that differ greatly from the one by Gustav III — the dilemmas posed by coffee and the effects it may have on our health. The most recent results seem to attest to the fact, at least according to some interpretations of observed data, that coffee does have a protective action against pathologies such as Alzheimer's and Parkinson's, whilst there is still a debate on whether coffee, especially when not filtered during the preparation, leads to an increase in cholesterol (Thelle and Strandhagen, 2005).

Who Invented Milk Chocolate?
And who discovered hot water?

The caption under the portrait of British naturalist and physician Hans Sloane (1660–1753), on display nowadays at the Royal Society — of which Sloane was the President after Isaac Newton — presents him as the owner of the original collection of the British Museum and as the inventor of milk chocolate. To this day, there are chocolate producing companies that make reference to his name ("Sir Hans Sloane *chocolatier*").

In 1687 Sloane left for a long trip around various islands, one of which was Jamaica. His medical training made him particularly interested in plants and their possible use in the food and therapeutic fields. He gathered hundreds of samples, plant drawings by artists and numerous

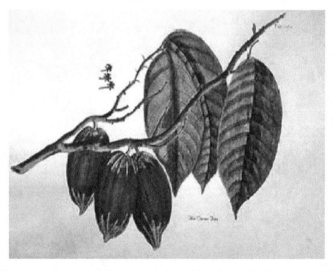

Fig. 3.1 Cocoa tree in an etching commissioned by Sloane (original image in Sloane, 1707–1725, II, tab. 160).

notes on the life of the local inhabitants of the island (mostly slaves imported from Africa to work on the sugar plantations) that he describes in two volumes of the *Voyage to the Islands Madera, Barbados, Nieves, St Christophers and Jamaica, with the Natural History of the Herbs and Trees, Four-footed Beasts, Fishes, Birds, Insects, Reptiles etc. of the last of those Islands*, more briefly known as *Natural History of Jamaica* (1707–1725).

Sloane brought a sample of a cocoa plant back from Jamaica, which is still preserved today at the Natural History Museum in London. Sloane notes in his writings that the locals "all drink chocolate, at all hours, but especially in the morning". After tasting it, he found it "nauseous, and hard of digestion, which I suppose came from its great oiliness, and therefore I was unwilling to allow weak stomachs the use of it, though children and infants drink it here, as commonly as in England they feed on Milk… The common use of this, by all people in several countries in America, proves sufficiently its being a wholesome food. The dinking of it actually warm, may make it the more Stomachic (easier to digest), for we know by anatomical preparations, that the tone of the fibres are strengthened by dipping the stomach in hot water, and that hot liquors will dissolve what cold will leave unaffected" (Sloane, 1707–1725, I, p. XX). Sloane started to prescribe it, especially to dilute medicinal plants, such as those with laxative properties for example.

To make chocolate more palatable to European taste, Sloane had the intuition of using milk instead of water. He became passionate about the drink, and his vast collection of objects includes numerous cups and containers for chocolate. His "recipe", copied and commercialised, became very successful and a popular saying continued to identify him as the inventor of hot milk chocolate and even as a successful entrepreneur in this sector. It is more probable that other entrepreneurs used his name and prestige, such as the owners of a brand who had a furious row with another company that used the name of Sloane for its own milk chocolate in 1775.

More significantly, the interest that Sloane had for the chocolate drank by the Jamaicans singularly intertwined "botanical and culinary narratives" with "collection and confection", medical practice and ethnographic observation. The addition of milk to the sour and nauseating indigenous beverage marked the difference between tastes yet also a form of inclusion and comprehension of the exotic. To the point that "chocolate" became a simplifying expression and jargon for a racial point of view "they are a kind of chocolate colour" wrote Thomas Aubrey in 1729 when he described a shipment of slaves (cit. in Delbourgo, 2011, p. 92).

It seems questionable however that Sloane was the first one to invent milk chocolate. The Spanish, who had derived the word Chocolate from a Nahuatl word meaning "sour water", had in fact introduced Cocoa to Jamaica. And chocolate had been circulating in Europe in various preparations since the second half of the 17th century. *The Indian Nectar, or a discourse concerning Chocolata,* was written in 1662 by the English physician Henry Stubbe, who observed at the time how "in England we are not content with the plain Spanish way of mixing Chocolata with water [... we] either use milk alone; or half milk, and half conduit-water" (Stubbe, cit. in Delbourgo, 2011, p. 78). In 1686, a year before Sloane's departure, the inventor of the pressure cooker Denis Papin (1647–1712) had tried to boil chocolate *in vacuo* during an experiment at the Royal Society, but the experiment had not been particularly successful, having not proved to be significantly different from traditional boiling methods.

In 1680 the microscopist and spice merchant Diacinto Cestoni asked his friend and colleague Francesco Redi for the recipe for "jasmine chocolate". His friend's reply was evasive to say the least, as if this was a discovery of which he had to protect the priority.

> I am sorry, dear Sir, that you should ask me something of which I have strict orders not to talk of: how to modify chocolate with the flavour of Jasmine. What I can tell you is that you do not achieve this with the water of the Jasmine, because cocoa will not blend with the water, and although one can add a few drops of scented water, it would not be enough to flavour the entire mass of chocolate with Jasmine. And if one used too much water, it would not combine with the chocolate. I know that you Sir are a discreet person, and therefore understand there are limits to what can be revealed. (Redi, *Opere*, 1809–1811, II, p. 47)[40]

---

[40] On Redi and Chocolate, see also Bernardi (2005).

As with coffee, during the following decades there were contrasting opinions on the effects of chocolate. In his *Parere intorno all'uso della cioccolata* (*Advice on the use of chocolate*) in 1728, Giovanni Battista Felici deplored the lightness with which with "no reason" or "experience" qualities such as relief from fever were attributed to this drink "from the coarse inhabitants of the Indies". Palpitations, agitation, laxative effects; "it seems in short that chocolate never leaves anyone who drinks it in their natural state of rest" (Felici, 1728, p. 9). Therefore based on a classic scheme, the new element from far away comes in to disrupt a "natural" order of things.

An answer published that same year, *Altro parere intorno alla natura, e all'uso della cioccolata*, (*Another Opinion on the Nature and Use of Chocolate*) presents a symmetrical argumentative strategy, underlining the affinities between the new drink and other already familiar ones. According to the author, by heating chocolate one obtains a spirit that is "innocent and milky, that combined with the aqueous proportion (...) and in virtue of all its examined components, I believe that one cannot disagree with those who consider the nature of Cocoa not much inferior to that of Milk" (Zeti, 1728, p. 17). And even if one does attribute to cocoa characteristics that are too strong or unusual, "by correcting it via Aromatic Drugs, the composition of Milk Chocolate becomes tempered and benign" (Ibid., p. 19). Hence like in the case of Sloane, by varying the ingredients and preparation, the exotic drink becomes "domesticated" and in line with habitual taste and guidelines.

In a similar context, Merton would have probably concluded that the "invention of hot milk chocolate" was mature and even inevitable, and it does not come as a surprise that many minds in those times — including Sloane — chose to move in similar directions.

Even Merton however would have been surprised by a debate on the "discovery of hot water", since this is a common expression used to mock someone who boasts about having discovered something well known and obvious.[41] But "if we don't know when, where or by whom hot water was invented, we can establish with no doubt that the

---

[41] As noted by Ceresa, in English and German speaking countries, there are similar expressions that refer to technology such as "reinventing the wheel". In France, the

invention of the science of hot water took place during the Tang dynasty in China (618–906), after the habit of drinking hot tea had spread to the whole country" (Ceresa, 1993, p. 10). An extensive production of over a hundred treaties on tea, from then on and during the following centuries starting from *Canone del te'* (*The Classic of Tea* — Chajing, 760 circa) by Lu Yu, classifies and analyses in detail the quality of the different waters (from springs, waterfalls and different rivers) and stages of ebullition with descriptions "that pre-empt those of modern physics" (Ibid., p. 12).

> (The boiling) with tiny bubbles like fish eyes and a light high pitched sound is considered the first boiling. The moment the bubbles, like a string of pearls, gather along the edge of the recipient, like in a rippling spring, is considered the second boiling. The one similar to breakers rising and waves breaking is considered to be the third boiling. Once the third boiling is over, the water grows old and can no longer be drunk. (Lu Yu, *Chajing*, cit. in Ceresa, 1993, pp. 31–32)

The link between this "science" and the preparation of beverages is very different from the one we have observed until this juncture: analytic observation, practical knowledge, sobriety and balance meet, to the point that "the discovery of hot water becomes the symbol of research of the essential elevated to perfection" (Ibid., p. 13).

---

expression goes "he didn't invent the wire to cut butter" which is an interesting culinary contextualisation (1993, p. 10).

Freezing Water, a Piece of Rubber,
Richard Feynman and the press (stir well, serve chilled)

This tragic story starts on a cold morning in January and culminates on 11 February 1986 in a glass of ice cold water. In a room packed with cameras and journalists, the physicist and Nobel Prize laureate Richard Feynman sits at a table. Feynman asks that someone bring him a glass of water with a lot of ice. Everyone expects him to drink it, but instead Feynman takes a piece of rubber out of his pocket and puts it next to the glass. He shows the piece of rubber to the journalists and squashes it between his fingers, to underline its elasticity. Then he suddenly plunges it into ice water. Once, and twice. He then takes out the piece of rubber and squashes it once again. The piece of rubber does not return to its original shape and breaks easily. The room and the photographers' flashes erupt into a loud buzz.

Flashback. Two weeks earlier, 28 January 1986. At 11.38am, seventy-three seconds after takeoff, the space shuttle Challenger explodes live on television, causing the death of the seven crew members including the teacher Christa McAuliffe. The impact on public opinion is huge. The President of the United States, Ronald Reagan, nominates a presidential commission, with Feynman as one of the members, to investigate the cause of the accident. It is during a public hearing that Feynman makes his demonstration with a piece of rubber and a glass of freezing cold water.

On 11 February, in front of the press, Feynman explains that the piece of rubber is made of the very same material as the O-rings meant to prevent the heated gases from escaping joints that connect the single segments of the rocket. The low reactivity of the material at cold temperatures (like those registered on the morning of the launch), demonstrated by immersing a piece into freezing water, had been the cause of the Challenger's explosion.

Top NASA officers and technicians from Morton Thiokol, the company which produced the O-rings, tried in vain to point out the differences between the "trick" with freezing water carried out in few seconds by the physicist in front of the cameras, without measuring the temperature of the water or taking into account other parameters, and the situation that had occurred during the launch of the Challenger. Feynman himself later admitted, "this is not the way to do such experiments" (Cit. in Gieryn and Figert, 1990, p. 77).

Newspaper headlines the following day titled: *Feynman explains the causes of the Challenger tragedy* and two years later, on Feynman's death, all the articles about him mentioned the "experiment that 'demonstrated' something about why *Challenger* exploded" (Ibid., p. 71).

Fig. 3.2 11 February 1989, Richard Feynman demonstrated in front of journalists the cause of the Challenger's explosion (Nasaspaceflight.com).

The charisma, the authority and public visibility of the Nobel laureate, combined with the narrative demands from the media, accomplished a miracle: a piece of rubber in ice cold water in front of the cameras became a decisive *Experimentum Crucis*. "This was not a piece of rubber with which to amuse oneself, but the joint of the Challenger squashed between the segments of the rocket (...). This was not a paper cup full of ice-cold water, but the cold temperatures of the day of the launch at Cape Canaveral (...). Feynman assembled the "correct ingredients", the same ones with which we can "cook up a theory of science in society", to understand the dynamics of credibility and trust in science in public" (Ibid., pp. 89–91).

The use of such a common object, of such a widespread drink, emphasises the basic character and the immediate understandability of the demonstration. The cause of the Challenger's explosion is there, under everyone's eyes. As simple as drinking a glass of water.[42]

---

[42] In other cases, the states and the qualifications of water are used metaphorically to criticise results of poor relevance or those that are not even considered part of the scientific domain. The controversial story of the so-called "polywater" (a polymerized form of water hypothesized in the late sixties, and later dismissed) was presented as nothing more than "dirty water" and those who criticised it were named the "impurity lobby" (Franks, 1981).

DESSERT

A Taste of Science (and Society)
From Brillat-Savarin to Molecular Cooking
via Futuristic Cooking

"There is something about human
beings: they don't just survive,
they discover, they create!"

REMY, *Ratatouille*

Culinary Science According to Brillat-Savarin

"The discovery of a new dish does more for human happiness than the discovery of a star." It is with this sort of aphorism, in the opening pages of his *Physiology of taste* (1825), that the magistrate but also chemistry and medicine enthusiast Anthelme Brillat-Savarin (1755–1826) claims the right for culinary themes to sit at the same table as topics generally treated by science. "Are we not surrounded by *gourmets*, who can tell the latitude in which any wine ripened just as surely as one of Biot's and Arago's disciples can predict an eclipse?" (Brillat-Savarin, 1825, Eng. trans. 1854, p. 71).

And quoting a line by Henrion de Pansey addressed to "three of today's best scientists (Messrs Laplace, Chaptal and Berthollet): "I shall never think science sufficiently honoured until I see a cook in the first class of the Institute" (Ibid., p. 346). Gastronomy claims its place next to the other sciences, "its sisters" that must recognise its importance as a source of sustenance, pleasure and sociability; it must assert itself and fight its way following the example of other "new sciences" such as "stereotomy, descriptive geometry and the chemistry of gas". At the same time, the study of food and its preparation should emerge from the pre-scientific stage and promote the ample research that is now available; "as long as the secret was kept in cellars, and where dispensaries were written, the results were but the products of an art" (Ibid., p. 75).

As we can see, other than the change in content that Brillat-Savarin wishes for, he would like above all for science to become an institutional model for food preparation in terms of subjects and the ways knowledge can circulate. At the same time, he claims dignity and scientific status for gastronomy and its experts: a skillful cook can be a "scientist in theory (and) in practice", placing himself "for the nature of his functions (...) between the chemist and the physicist" (Ibid., p. 284).

As the "scientific definition of all that relates to a man as a feeding animal", "gastronomy is a chapter of natural history, for the fact that it makes a classification of alimentary substances. Of physics, for it examines their properties and qualities. Of chemistry, from the various analysis and decomposition to which it subjects them. Of cookery, from the fact that it prepares food and makes it agreeable. Of commerce (…) and lastly of political economy" (Ibid., p. 77).

This assimilation to the forms and languages of science takes on picturesque elements when it debates "frying theories", or when, while discussing great appetites the author remembers a guest who "in a large white cheese" cut a "ninety degree angle breach"; not to mention the description of a boiled food based on the model of a scientific manual.

> A piece of beef, intended to be cooked in boiling water, slightly salted so as to extract all the soluble parts. Bouillon is the fluid which remains after the operation. Bouilli is the flesh after it has undergone the operation. (Ibid., p. 102)

Another element of "scientificity" is identified by Brillat-Savarin when he collects and compares various points of view, with observations that often have analogies with the judicial field with which he is professionally familiar with. After having taken note of the confession of a lady who claims the effect of truffle caused her to have an encouraging attitude towards an admirer, he claims that

> A confession, no matter how sincere can never give origin to a scientific law. (…) I made ulterior researches, collected my ideas, and consulted the men who (thanks to their social status, receive the most confidences) I united them into a tribunal, a senate, a Sanhedrim, an areopagus, and we gave the following decision (…): the truffle is a positive aphrodisiac, and under certain circumstances makes women kinder, and men more amiable". (Ibid., p. 127)

Conclusions and definitions that, in reference to a developing model of scientific discussion and debate, can only be provisional: "In the present state of science we understand by sugar a substance sweet to the taste, crystalizable, and which by fermentation resolves itself into carbonic acid and alcohol" (Ibid., p. 181).

The different camps in terms of gastronomy are almost a caricature of the most violent scientific controversies between the supporters of alternative theories and the debates that oppose supporters and critics of a new result or interpretation:

> "Master La Planche", said the professor with that deep grave accent which penetrates the very depth of our hearts, "all who sit at my table pronounce your potages of the first class, a very excellent thing, for potage is the first consolation of an empty stomach. I am sorry to say through that you are uncertain as a *friturier*. I heard you sigh yesterday over that magnificent sole you served to us, pale, watery and colourless. (…) This happened because you neglected the theory, the importance of which you are aware of. You are rather obstinate, though I have taken the trouble to impress on you the facts, that the operations of your laboratory are only the execution of the eternal laws of nature, and that certain things which you do carelessly, because you have seen others do so; yet these are the results of the highest science."
> (Ibid., p. 158)

Similarly, others were inspired by the model of scientific experiment, which Brillat-Savarin calls "gastronomic trials" (*éprouvettes gastronomiques*): "dishes of so delicious a flavour that their very appearance excites the gustatory organs of every healthy man. The consequence is that all those who do not evince desire, and the radiancy of extasy, may very properly be set down as unworthy of the honours of the society and the pleasures attached to them" (Ibid., p. 198).

It should be noted that the logic of these trials was reversed compared to the usual models in which one experimented with the effect of a substance: here the effect is known, and what is examined is the reaction of the subject to whom the substance is administrated. It is therefore the subject who is classified (as "worthy or unworthy", in the author's terminology, and as a gourmet or not) on the basis of his reactions.

This is an experimental model that is most similar to the reactions of an antibody on the basis of which a subject is classified as "positive" or "negative". In science studies that analyse experimental practices, one talks of a "range of results" that are considered acceptable and therefore allow checking the accuracy of detection instrument or measurement (Collins and Pinch 1993; Bucchi, 2004).

These trials must furthermore be adapted to the social provenance of the subjects "it's a dynameter the power of which should increase as we ascend in society. The test for a householder in Rue Coquenard would not suit a second clerk and would be unnoticed at the table of a financier or a minister" (Brillat-Savarin, quote, p. 198).

Another precaution to be considered is relative to the "scope of the experiment": "to ensure that a trial produces the effect, it is necessary that it be used on a wide scale: experience, based on the knowledge of mankind has taught us that the most tasty rarity loses its efficiency if it does not appear in large quantities: because *the first reaction it triggers in guests* is prevented by the fear of receiving too small a portion or to be, in certain situations, obliged to decline out of courtesy: which is often the case in the homes of ostentatious misers" (Ibid., p. 169, my italics).

Brillat-Savarin also makes reference to the equally interesting proposal of "negative and depravation trials. For example, one could image that some misfortunate event had ruined a delicious dish, or a basket that had meant to arrive by post had been delayed (...), this catastrophic news would have caused a gradual sadness on the face of the guests and one could have thus obtained a good measure of their gastric sensibility" (Ibid., pp. 167–168).

The objection to these "negative trials" is extremely modern, because their nature is not strictly methodical, but rather "ethical" as we would say nowadays: "Similar events, which would superficially impact the poor organs of the indifferent, could cause to the true believers a fatal influx and possibly a mortal stroke. Therefore notwithstanding a few insistences from the author, the proposal was discarded unanimously" (Ibid., p. 168).

There is even a hint to the psychology of knowledge, when the author for example asks himself what has encouraged Linnaeus to assign to cocoa the name of *cacao theobroma* (the food of the gods). An "emphatic title" of which some attribute the cause to "his passionate fondness for it, and the other his desire to please his confessor; there are those who attribute it to gallantry, a Queen having first introduced it" (Ibid., p. 145).

But above all, in the *Physiology of taste*, one already glimpses with clarity that discursive model through which science elevates common sense to a new "culinary awareness". When a concept needs to be defined, the "popular answer" is immediately followed the more detailed and analytic "scientific answer".

> What is understood by aliments?
> Popular answer: All that nourishes us
> Scientific answer. By aliments we intend the substances which, submitted to the stomach, may be assimilated by digestion and repair the losses which the human body is subjected to in life. (Ibid., p. 91)

In the concluding "Envoy to the Gastronomes of the Two Worlds", the projection by Brillat-Savarin goes once again in the direction of an institutional model of science: as in other scientific activities, a fruitful encounter can take place between — in a fashion typical of the culinary field — the need for recognition and satisfaction of the single person and the institutional need for the knowledge development.

> Work, your Excellencies, work for the good of science: digest for your personal interest: and if, during the course of your work, you come across an interesting discovery, do communicate it to the most humble of your servants. (Ibid., p. 382)

Science in the Kitchen According to Pellegrino Artusi
Or the importance of scientific celebrities

Fig. 4.1 P. Artusi, *La scienza in cucina e l'arte di mangiar bene*, 1891.

"Celebrity Scientists": scholars in science communication thus refer to those scientists, among whom Nobel prize winners often surpass, who have great public visibility and the capacity to influence not just their own colleagues but also the media and the wider public.

The founder of science sociology Robert K. Merton had already grasped the scope and relevance of this phenomenon, and its cumulative effect. He called it the "Matthew Effect" based on the passage of the gospel that states, "For unto every one that hath shall be given, and he shall have abundance: but from him that hath not shall be taken away even that which he hath" (Matthew 25, 29). Those in positions of visibility and prestige will have privileged access to further resources and positions of visibility, and so on. A scientific result will have greater visibility in the community of scientists when it is introduced by a scientist of high rank than when it is introduced by one who has not yet made his mark (Merton, 1973, p. 473). Or as a Nobel prize-winner for physics put it: 'The world is peculiar in this matter of how it gives credit. It tends to give the credit to [already] famous people' (Merton, 1973: 473). On analysing empirical data, Merton discovered, for example, that papers submitted to a scientific journal were accepted more frequently if one of the authors was a Nobel prize-winner or a particularly well-known researcher. Likewise, papers by a scientist were cited much more frequently after that person had received some highly visible award like the Nobel Prize.

Today the Matthew effect is accentuated and amplified under the pressure of public relations offices and the frequent short-circuits between the research sector and media communications. Thus, science becomes at least partly permeated by a "star system" dynamic, not so different from the one characterising the world of media or sport. Scientists become familiar to the general public (the case of Nobel prize-winners is emblematic) as "brands" that are marketable in the most disparate sectors.

The media are interested not only (or not so much) in the actual research results by science superstars like the late Stephen Hawking or Craig Venter, but also in their political opinions and their love relationships, just as with other celebrities (Beltrame, 2007; Fahy and Lewenstein, 2014). Some of them, such as Richard Dawkins, owe most of their notoriety and capacity to influence the public debate to the success of their popularisation activities more than to any original contribution to research.

Although they were not as strong at the time, these dynamics already had some form of relevance at the end of the XIXth century. To benefit from the recommendation or even from the "sponsoring" of a visible scientist and to be able to "piggyback" on their notoriety could contribute to acquiring visibility and relevance.

This is what happened to Pellegrino Artusi and his work *La Scienza in cucina e l'Arte di mangiar bene (Science in the Kitchen and the Art of Eating Well*, 1891). Based on what the author himself recounts in the opening pages, of the "story of a book that is a bit like the story of Cinderella", the text received scathing reviews and was turned down by numerous editors. Printed at the author's expense, it did not succeed in finding its way into the heart of readers, nor in triggering any interest in the press — Artusi recalled with bitter disappointment that a magazine to which he had sent the text had "not accorded (it a) few words of praise (...) it was actually listed in the rubric of books received, with a mistake in the title! (Artusi, 1891; Eng. trans. 2003, p. 1).

Until the day, in the place of a fairy godmother, a celebrity scientist sponsors him.

> Finally, after so many setbacks a man of genius suddenly appeared and took up my cause. Professor Paolo Mantegazza, with that quick, and ready wit that is his trademark, immediately recognised that my work indeed had some merit, and might be of use to families. Congratulating me for my work, he said: "With this book, you have done a good deed; may it have a thousand editions. (Ibid., p. 58)

"A picturesque figure" — this is how the medical journal The Lancet remembered Mantegazza after his death — a physician, holder of the first chair of anthropology in Italy, a senator, a prolific author of popular books of huge success both in Italy and abroad, a novelist, he was a figure of great authoritativeness and notoriety. Among his most successful publications are the so-called "physiologies", starting with the *Fisiologia del Piacere* (*Physiology of Pleasure* 1854, re-printed many times and sold in tens of thousands of copies), a testimony to the fact that in Italy too, the term physiology had by then "escaped the traditional use and had become part of the fashionable scientific jargon"[43] as seen previously in the works of Brillat-Savarin.

---

[43] Govoni (2002, p. 219). A lecturer presenting a "physiology of credit" observed ironically: "love, pleasure, pain, the soul, thoughts, society, every sin (...) had their physiology. And tomorrow (...) there will be a physiology of nothingness or the physiology of the slacker!. So why not a physiology of credit?" (Ibid., p. 219). On Mantegazza's vision of science popularisation see also Turbil (2017).

In this context, cooking and the preparation of foods emerged as a great didactic and educational opportunity to introduce the greater public to science and conquer it in an almost "missionary" way. Oscar Giacchi in *Il Medico in Cucina* (*A Doctor in the Kitchen*, 1882) wrote with a significant religious analogy that "if the dogmas of science are such truths that they impose adoration and silence (on us) or we risk the highest level of excommunication, then physiology, like all religions, needs its popular catechism to explain to the ignorant the reasons for its cult and its rites; and to teach how to make the most of its beneficial teachings" (Giacchi, 1882, p. 12).

Mantegazza himself had taken an interest in cooking, for example in his *Igiene della Cucina* (*The Hygiene of Cooking*), one of the small volumes of the *Enciclopedia Igienica Popolare* (*Encyclopaedia of Popular Hygiene*), of which the title page has a quote by Brillat-Savarin: "Tell me what kind of food you eat, and I will tell you what kind of man you are": a series of notes and advice on the nutrition and digestive values of various foods. In the book by Artusi, he most likely saw a contribution to the programme of uniting "science, hygiene and morality" based on which "that which is healthy is honest, that which is true (scientifically) is also moral" (Govoni, 2002, p. 257). Such a programme of lay moralisation founded on the scientific knowledge of the body of the nation, this "Italian pastoral" — to paraphrase Philip Roth — found a concrete expression and even enrichment in the work by Artusi. The book outlines from the title "where science, through the mediation of art, becomes practical (...), a culinary triangle of culture, invention and experience that finds an identical equivalent in that other triangle made up of *hygiene* (science), *economics* (practice), *good taste* (art)". As such "the proposition of science in the kitchen becomes a commitment that is positivist at a time when scientism was very widespread: it is the language of the cultural journalism of those times that does not shrink from the attraction of scientific positivity". Artusi furthermore extends Mantegazza's equation to another term, "good taste", that amplifies and reinforces its value in the culinary realm (Camporesi, introduction to Artusi, 2001, p. XXXI).

A great admirer of Mantegazza, in the following editions, Artusi would boast not only the wishes bestowed upon him by the physician and populariser, but also a letter of congratulations from his widow, countess Maria Fantoni ("I have made your quince jelly, and it is now on its way to America: I sent it to my stepson in Buenos Aires (...) you write and describe things so clearly that executing your recipes is a true pleasure and brings me real satisfaction" (Artusi, cit., p. 23).

It is well known that there is little science, strictly speaking, in Artusi's book, although there is no lack of caption notes of a naturalistic nature at the beginning of some of the recipes, thus Artusi seems to have been quite inspired by the experience of his friend Enrico Hillyer Giglioli, a naturalist and director of the Zoological Museum of Florence. This is for example the case for his Stewed Angel Shark ("Angel Shark, has a flat body similar to the ray", Ibid., p. 331), for the long description that precedes the eel based recipes ("recent studies in the strait of Messina have shown that this fish and those of the family of the eel, need to lay their eggs in deep sea at a depth of over 500 metres and that, similarly to frogs, they undergo a metamorphosis", Ibid., p. 276) and even more so that of the sturgeon.

> I hope my readers will allow me to give a little history on this very interesting fish. The sturgeon belongs to the order of the *Ganoidei*, from the Latin *Ganus,* which means shiny, owing to the shine of its scales, and to the sub-order of the *Chondrostei* since it has a cartilaginous skeleton. It constitutes the family of the *Acipenser* which is defined by those two characteristic qualities, as well as by a skin made up of five longitudinal series of shiny scales. (Ibid., p. 341)

The entry on chocolate offers him an opportunity to quote the eminent Mantegazza.

> Like other foods that stimulate the nervous system, chocolate also excites the intellectual faculties and enhances sensitivity; but, as it is rich in albumin and fat (cocoa butter), chocolate is also very nourishing, acts as an aphrodisiac, and is not easily digested. (Ibid., p. 77)

And when speaking of coffee, Artusi even allows himself to contradict his mentor, albeit gently.

If coffee causes too much agitation and insomnia, it is better to abstain from it or use it in moderation (...) Coffee seems to cause less agitation in humid, damp places and perhaps this is the reason the European countries where the consumption of coffee is highest are Belgium and Holland. (...) As for what Professor Mantegazza says, that is, that coffee does not in any way aid in digestion, I believe that it is necessary to make a distinction. He perhaps would say that this is true for those whose nervous system is indifferent to coffee; but for those whose nervous system (including the pneumogastric nerve) is affected by this beverage, it is undeniable that they digest better after drinking it, and the prevailing custom of having a good cup of coffee after a rich meal confirms this. (Ibid., p. 71)

"Abbasso la Pastasciutta" (Enough with Pasta)!
Science and technology in futurist cooking

On 28 December 1930, the *Manifesto della Cucina Futurista* (*Manifesto of Futurist Cooking*) by Filippo Tommaso Marinetti is published in Turin's *Gazzetta del Popolo*. The manifesto's objective is "to renew totally the Italian way of eating". The nutritional habits of the Italians are considered to be out of touch with the rapidly changing times and particularly with their new technologies.

> We futurists feel (...) that we must stop the Italian male from becoming a solid leaden block of blind and opaque density. Instead he should harmonize more and more with the Italian female, a swift spiralling transparency of passion, tenderness, light, will, vitality, heroic constancy. Let us make our Italian bodies agile, ready for the featherweight aluminium trains which will replace the present heavy ones of wood iron steel. (Marinetti and Fillìa, 1932, Eng. trans. 1989, p. 36)

The first point in the programme proposes "The abolition of pastasciutta, an absurd Italian gastronomic religion. (...) it is completely hostile to the vivacious spirit and passionate, generous, intuitive soul of the Neapolitans. (...) When they eat it they develop that typical ironic and sentimental scepticism which can often cut short their enthusiasm." (Ibid., p. 37.) They also do not refrain from citing, in support of their aims, expert advice such as that by

> "(...) highly intelligent Neapolitan Professor, Signorelli, writes: 'In contrast to bread and rice, pasta is a food which is swallowed, not masticated. Such starchy food should mainly be digested in the mouth by the saliva but in this case the task of transformation is carried out by the pancreas and the liver. This leads to an

interrupted equilibrium in these organs. From such disturbances derive lassitude, pessimism, nostalgic inactivity and neutralism". (Ibid., p. 37)

Pastasciutta is synonymous here of an outdated society and technique, of "tangled threads and somnolent old sailing-ships"; the "invitation to chemistry" that the Manifesto outlines in the area of gastronomy is necessary to project into a future that is already coming true, shaped by modern technologies such as the transmission of radio signals.

> Pastasciutta, 40% less nutritious than meat, fish or pulses, ties today's Italians with its tangled threads to Penelope's slow looms and to somnolent old sailing-ships in search of the wind. Why let its massive heaviness interfere with the immense network of short long waves which Italian genius has thrown across oceans and continents? Why let it block the path of those landscapes of colour form sound which circumnavigate the world thanks to radio and television? (Ibid., p. 37)

Whilst according to Mantegazza and Artusi the domestic household used to be the privileged context for administering scientific knowledge though food, it is now the State that should take on this responsibility through a combined programme of science, gastronomy, politics and economics.

> We invite chemistry immediately to take on the task of providing the body with its necessary calories through equivalent nutrients provided free by the State, in powder or pills, albumoid compounds, synthetic fats and vitamins. This way we will achieve a real lowering of the cost of living and of salaries, with a relative reduction in working hours. Today only one workman is needed for two thousand kilowatts. Soon machines will constitute an obedient proletariat of iron steel aluminium at the service of men who are almost totally relieved of manual work. (Ibid., p. 38)

Here Futurism, in its characteristic fashion, outlines an analogy for the entry of science into cooking that stretches back centuries, continuing until contemporary molecular cuisine: modern "civilised" cooking is associated with lightness and dematerialisation of food, symbolised by the soufflé.

> After powdered food, we will see food made of gas, and nutritional ether (...) we turn everything into mash for civilisation. Therefore we shall make gas-based delicacies... for now we shall cook soufflés. (Pettini, 1905, p. 78)

Ignorance and past, pre-scientific forms of cooking are associated with the 'opaque and blind' heaviness of pasta; science in the kitchen promises agility, "passion and spiral transparency". Among the precepts of futurist cooking, there are also a series of 'scientific instruments for the kitchen', with which to transform foods and organise previously approximate cooking methods and systems:

> (...) ozonizers to give liquids and foods the perfume of ozone, ultra-violet ray lamps (since many foods when irradiated with ultra-violet rays acquire active properties, become more assimilable, preventing rickets in young children, etc.), electrolyzers to decompose juices and extracts, etc. in such a way as to obtain from a known product a new product with new properties, colloidal mills to pulverize flours, dried fruits, drugs, etc.; atmospheric and vacuum stills, centrifugal autoclaves, dialysers. The use of these appliances will have to be scientific, avoiding the typical error of cooking foods under steam pressure, which provokes the destruction of active substances (vitamins etc.) because of the high temperatures. Chemical indicators will take into account the acidity and alkalinity of the sauces and serve to correct possible errors: too little salt, too much vinegar, too much pepper or too much sugar. (Marinetti and Fillìa, cit., p. 40)

Science and technology are thus called upon to emancipate cooking from ignorance and approximation, and to project it towards a new and modern concept of "good taste" which encapsulates scientific knowledge, rational production, physical efficiency, nationalistic volition and aesthetic satisfaction.

The head reporter at the *Gazzetta del Popolo* agrees with Marinetti when he declares that "when encumbering and soporific pasta is banned from the tables of the peninsula, when the kitchen is no longer the dominion of inept housewives and ignorant, poisoning cooks, but becomes a source of wise chemical combinations and aesthetic sensations, when we succeed in creating and diffusing a way of eating that can reconcile tiny quantities with maximum explosive dynamic nutritive power, only then will the will-power, vitality, imagination and creative genius of the race reach its apogee" (Ibid., p. 47, italics mine).

As one can imagine, the futurist manifesto triggers several controversial reactions. A few scholars such as Bettazzi, Foà, Pini, Lombroso, Ducceschi, Londono, Viale leap to the defence of pasta. But futurists swiftly dismiss their point of view and question their scientific legitimacy. "Most un-scientifically they obeyed the dictates of their palates. They seemed to be speaking from a table in some trattoria in Posillipo, with their mouths blissfully full of spaghetti alle vongole. They do not have the spiritual lucidity of the laboratory. They forget the lofty dynamic obligations of the race and the searing speed and most violent contradictory forces that constitute the agonizing rush of modern life (...) All the defenders of pasta and implacable enemies of Futuristic cooking are melancholy types, content with their melancholy and propagandists of melancholy" (Ibid., p. 41).

There follows a list of 'expert' opinions against pasta and in support of Marinetti's programme. According to the clinician Pende "the common, and exaggerated consumption of pasta leads to inevitable weight gain and to exaggerated abdominal volume. Great pasta consumers have slow and peaceful temperaments, those who eat meat are fast and aggressive"; according to the clinician and senator Gabbi "one should change eating habits to follow the laws of biology; the constant repetition of the same food has been proved to be harmful".

During the following years, the Manifesto inspires dishes such as the complex edible plastic Equator + North Pole created by the futurist painter Enrico Prampolini, i.e. "a cone of solidified whipped egg whites with orange quarters shaped like sun rays. The tip of the cone is covered with pieces of black truffle cut into dark airplane shapes, ready to conquer the zenith". Another menu was the "Aeromeal", invented by Marinetti himself, and served at the Negrino Hotel in Chiavari in 1931. It includes a "start-up flan", a "palate-lift-off broth", "beef in fuselage" and "candied atmospheric electricity" for dessert. Another painter, Fillia (Luigi Colombo), with whom Marinetti dedicated an entire volume to futurist cooking, invented a multisensory dish called "the aerodish", in which food was accompanied by strips of velvet, satin and sandpaper, to be brushed with one's right hand in order to achieve pre-labial sensations thus rendering the food mouth wateringly good". The trattoria *Santopalato* ("Saint Palate") in Turin, served the "pollofiat" ("Chickenfiat"), stuffed with ball bearings in order for the meat to absorb the sweet taste of steel (cit. in Beccaria, 2009, pp. 16–17). The "futurist aeropainter" Caviglioni invented recipes such as "cosmic apparitions" (vegetables cut into star and moon shapes), "veal fuselage" and "rising turbot". Doctor Sirocofran is said to have invented the "formula for astronomic dinner", which was served in a pitch-dark dining room in crystal dishes, and which included "a consommé broth, rendered fluorescent with a minimal quantity of 'fluorescein', a cosmographic sphere of spumoni, and a telescopic pump which sprayed Asti sparkling wine".

In a page dedicated to "Christmas futurist cooking", the artist Fortunato Depero praises Marinetti and Fillia's book and adapts it using geometric terms, suggesting and sketching a "circle of juices, fragrances and noises" which among other things included "lemon, orange and tangerine quarters; a noise-making roll to be chanted by a chorus, a small

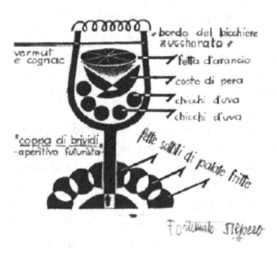

Fig. 4.2 Fortunato Depero, *Cucina futurista nataliza [Futuristic Christmas cooking]*, 1932.

rod of anchovies, candied chestnuts and dates". Depero also imagines a "household square", with "squares of polenta and crosses of green peas" and also "slices, slithers, stars, circles, triangles, squares, half-moons, hexagons, spires and zigzags of meat and raw, cooked or fried vegetables (Depero in *Il Brennero*, 25 December 1932, "Cucina futurista natalizia", illustrated).

The Birth of Molecular Cooking
Or how a meeting between a teacher and a physicist who dreamt of a winter equivalent of the ice-cream cone, brought scientists and cooks to the same table

In 1988 a cooking teacher from Berkeley, Elizabeth Cawdry Thomas, accompanied her husband to a convention at the "Ettore Majorana" centre in Erice where she had the opportunity to talk to a physicist from the University of Bologna, Ugo Valdrè, about the lack of attention given to the scientific foundations in the preparation of food. Valdrè encouraged her to follow up on the idea, and to organise a talk on "Science and Gastronomy" in Erice. The two of them contacted an Oxford physicist and fellow of the Royal Society, Nicholas Kurti, who was interested in the topic and who had led a programme on science and cooking on television in 1969, to ask if he might direct the workshop. Kurti accepted, and suggested they involve Harold McGee, a physicist and author of important works on the theme, including the famous *On Food and Cooking: The Science and Lore of the Kitchen*, and Hervé This, a chemist by training and the director of *Pour la Science*, the French edition of *Scientific American* ("and also a good cook", Kurti added). Kurti then prepared a list of possible guests: experts in haute cuisine such as Raymond Blanc (chef and owner of the Manoir aux Quat' Saisons near Oxford which had three Michelin stars at the time) and Guy Savoy of the eponymous restaurant in Paris: chemists, physicists and scholars in biomedicine among them Sir Arnold Burgen, Patrick Etievant (director of a "research laboratory on aromas"), Pierre-Gilles de Gennes of the College de France (author among other things of a speech in Oxford, Kurti observed, on "Foam, bubbles and other fragile objects — such as mousse, meringues, etc...").[44]

The Erice centre requested a title for the workshop that was "less frivolous" and more academic in view of printing the official programme: this led to the announcement of the first "International workshop on Molecular and Physical Gastronomy", that took place at Erice from the 8th to the 13th of August 1992. According to McGee, the term "molecular" was chosen in analogy with molecular biology, that was the "*hot scientific field*" at the time ("vivacious and emerging", but seeing as the theme was culinary, the choice of the word "hot" by McGee seems significant, as it could refer to "boiling").

In the flyer for the event, the objectives of the workshop were introduced by two quotes. The first was the already mentioned aphorism from the *Physiology of Taste* by Brillat-Savarin, according to which a new tasty dish in the kitchen is worth more than a new astronomical discovery. The second was from the American physicist Benjamin Thompson, count of Rumford (1753–1814) who dedicated many recipes and experiments to the preparation of food and invented an innovative coffee brewing machine. The quote from Thompson referred more specifically to the application of scientific knowledge to the preparation of food: "in which art or science could the obtained progress contribute most to the comfort and the pleasure of mankind?"

The Erice workshop proposed to answer the following questions:

a) To what extent do we understand the science that is at the base of culinary processes?
b) Could existing culinary methods be improved by a better comprehension of their scientific bases?
c) Could existing culinary processes improve the quality of the finished products or lead to innovations?

---

[44] This reconstruction is largely based on original documents (letters, promotional material and notes) made available by Harold McGee on his blog "The Curious Cook", www.curiouscook.com, consulted on December 15th 2011. The other quotes from McGee are from the same source, unless specified differently.

d) Could the processes developed in the framework of food production and large scale catering be adapted to the kitchens of houses or restaurants?

The programme of discussions was as follows:

Bouillon: concentrating by boiling at reduced pressure

Roux Sauces

Emulsion Sauces (mayonnaise, *hollandaise*)

Foams, meringues, mousses, soufflés

Doughs, breads, cakes

Milk and Cream: pasteurised vs. fresh. Viscosity, "whippability"

Cheeses: the role of micro-organisms

Making unedible plants edible

Microwave cooking

Deep Freezing. Food Irradiation

The perfusion of animal tissues. Marinating and tenderizing meat

Text and texture

Making safe and healthy eating compatible with good eating

When he received this programme, the physicist Valdrè recollected with pleasure their original encounter at Erice and complimented Cawdry Thomas. Because he could not attend the workshop, he asked her to put a problem to the participants on his behalf. He wanted to find a winter alternative to Ice-cream, that Valdrè thought he could call "Caldone" with the following requisites:

Caldone should fulfil the following requirements:

1. To be hot, as the name implies;
2. To be long lasting, i.e. something that can be licked;

3. One should be able to eat it during a walk (like an ice-cream, *gelato da passeggio*). It should therefore be placed in a container (possibly edible) of low thermal conductivity and good heat resistance;
4. Caldone should have a high thermal capacity, and be watery;
5. It should not leave a feeling of thirst afterwards;
6. It should be easy to prepare, easy to deliver on request and should be cheap;
7. It should be available over a long period of time during cold seasons;
8. It should have, of course, a good taste, slightly sweet.

As I was gradually describing Caldone to Nick (Kurti), he commented: "I have already made Caldone." He meant the opposite of "Baked Alaska". However, he agreed that it did not fulfil all the features of Caldone, in particular essential item 3. "Something hot to be eaten while walking is Caldarroste" (i.e. freshly roasted chestnuts), said Nick. Again requisite 5 is not fulfilled; in addition Caldarroste are not always available (req. 7). "I will think about Caldone; however, it seems to me possible to fulfil all requirements," concluded Nick.[45]

Other editions of the workshop were organised every two years until 2004. In 1998, after the death of Kurti, Hervè This dedicated the workshop to his memory and abbreviated its title to "molecular gastronomy".

According to Harold McGee, the Erice meetings "did not delve into cooking at the molecular level, *à la* molecular biology. Nor did they primarily emphasise innovation. The focus was on traditional kitchen preparations, how they work and how they might be improved by an understanding of the basic physics and chemistry involved. The idea that the workshops marked the birth of a new scientific discipline was never brought up in general discussion."

McGee does not even see a close link with the subsequent diffusion and public visibility of a "movement" of molecular gastronomy linked to

---

[45] Letter from Ugo Valdrè to Elizabeth Cawdry Thomas, 9 March 1992, cf. previous footnote.

some representatives of haute cuisine. "Of all the chefs who have come to be associated with 'molecular gastronomy', Pierre Gagnaire and Heston Blumenthal were the only ones to attend the Erice workshop. Gagnaire was entertainingly ambivalent about the idea of technical innovations in haute cuisine. Blumenthal came to the last two, in 2001 and 2004, but had already begun to apply science to restaurant cooking (…) Ferran Adrià, the most influential pioneer of experimental cooking, was never invited to the Erice workshop (…) Adrià had begun his programmatic pursuit of innovation in 1988, and set up a laboratory-workshop dedicated to research and innovation in 1997."

The point of view of Hervé This, the third co-director of the Erice meetings is slightly different; in an article — written for a journal of molecular biology — he did not hesitate to define molecular gastronomy as a new science, in which the term 'molecular' "has the same definition as it does in molecular biology. The similarity is intentional, because chemistry and physics are at the core of this discipline (…) for example, why a mayonnaise becomes firm or why a soufflé swells" (This, 2006, p. 1062).

This also admits there is a slight confusion when it comes to molecular gastronomy, especially in the media, a confusion that is at least in part due to the "mistakes Nicholas Kurti and I made when we created the discipline in 1988".

> Initially (…) molecular gastronomy had five aims; to collect and investigate old wives' tales about cooking; to model and scrutinise existing recipes; to introduce new tools, products and methods to cooking; to invent new dishes using knowledge from the previous three aims; and to use the appeal of food to promote science. Today it is easy to see that this scientific program was misleading and had shortcomings (…) The first two objectives are really scientific goals, the third and fourth are only technological applications, and the fifth aim is an educational application of the first four. (Ibid., p. 1064)

Is it not difficult at this point to see a continuity between the programme of molecular gastronomy and the long tradition that goes from Brillat-Savarin to the copious production of manuals on science and physics in cooking between the second half of the XIXth and the beginning of the XXth century, to futuristic cooking: to bring science into the kitchen, to update food preparation in line with new scientific knowledge and technological development. In this vision of molecular gastronomy, cooking is like an island that has remained oddly untouched in an otherwise civilised archipelago; like the village of Asterix that obstinately resists the Roman Conquest of Gaul. If every aspect of society has welcomed science, how can cooking still be immune to it?

> Despite having a huge impact on other aspects of our lives, scientific advances have done little to change our cooking habits. When it comes to preparing food (...) citizens in developed countries still cook almost in the same way as their ancestors did centuries ago (...). Kitchens are equipped with the same pans, whisks and sieves that cooks used in the XVIIth century. Similarly, culinary books from the XIXth to the XXIst century all look the same, despite the introduction of new recipes; for example, the first emulsion described in a French culinary book appeared in 1674 and the ancestor of mayonnaise seems to be a beurre de Provence. (Ibid., p. 1063)

The mayonnaise with which we started keeps recurring to define the relationship between science and cooking. It symbolises the delay and the ignorance in which common sense remains stuck with regard to a "non-scientific" preparation of foods: it goes back to the need to "incorporate" science to finally make cooking more modern and informed. Likewise, the rhetorical use of the female figure ("the old wives' stories") returns to symbolise the obstinate reluctance of common knowledge to abandon false myths and folkloristic superstitions and to embrace science, the only thing capable of revealing the otherwise mysterious change in state of mayonnaise.

> Indeed, cooking was the last of the "chemical arts" to become the object of scientific scrutiny and it is still relying on a tell-tale and anecdotal knowledge rather than solid science. As recently as 2001, an inspector from the French Department of Public Education said, during a public lecture that her mayonnaise failed when she had was menstruating. Such old wives' tales were

partly the reason behind the creation of molecular gastronomy: I first started experimental studies of cooking after encountering a recipe for cheese soufflé that advised adding egg yolks "two by two, never by fractions" (…). The late Nicholas Kurti, professor of physics Oxford University, UK, was upset by the poor and unscientific way that people cook. (Ibid., p. 1063)

The other preparation that is frequently mentioned is the soufflé, symbol of the impalpable "de-materialisation" and understanding that cooking can tend towards under the guidance of science. It becomes a new and wispy interpretation, in a vision that is strongly centred on science and the claim by Brillat-Savarin that "the discovery of a new dish confers more happiness to humanity than the discovery of a new star". Kurti viewed as a "paradoxical fact [...] that we know the temperature at the center of the planets and the sun better than the temperature at the heart of a soufflé" (This, 1993, Eng. trans. 2007, pp. 8–9).

Although This considers it inappropriate to call some chefs "molecular chefs" — "because chefs create food, not knowledge" — there is no doubt that culinary celebrities such as Ferran Adrià or Heston Blumenthal contributed to give weight to the term; an event that is often mentioned is a menu in "science and cooking" served by Pierre Gagnaire at the Académie des Sciences, during a conference on molecular gastronomy.

The modernisation and "scientificisation" of cooking are also expressed through the creation of innovative dishes that are given the names of famous chemists or physicists: *Gibbs* (an emulsion of eggs and oil, later cooked in a microwave, named after the physicist Josiah Willard Gibbs); *Vauquelin* (an egg white beaten and cooked in a microwave, inspired by the name of one of the mentors of Lavoisier, Nicolas Vauquelin), *Baumé* (an egg that is coagulated by immersing it in ethanol for approximately one month, named after the French chemist Antoine Baumé).

"This is one way of fighting the public's fear of science and of promoting the diffusion of knowledge" claimed This. The educational mission of molecular gastronomy, which takes us back to the initial television programme — of using the domestic and familiar character of the kitchen to draw the public closer to science — presents itself in the most direct and concrete way possible, by administering scientific information in the form of food. "This way people eat chemistry nowadays" the author concludes to his own satisfaction (2006, p. 1064).

Molecular Cooking
Science as an inspiration or as a "safe-haven investment"?

One of the most interesting aspects of the molecular gastronomy phenomenon is that it enables us to analyse the relationship between science and cooking, and more generally the one between science and society from various points of view. Therefore it is not only an opportunity for science to reach and colonise the territories of common belief such as cooking, but it is also a resource for cooking and for some of its most illustrious protagonists. In 2006, three of the most famous international chefs, Ferran Adrià, Heston Blumenthal, Thomas Keller, and the physicist and cooking expert Harold McGee presented a document to the press in which they defined "new cookery".

> Our cooking values tradition, builds on it, and along with tradition is part of the ongoing evolution of our craft. The world's culinary traditions are *collective, cumulative inventions*, a heritage created by hundreds of generations of cooks. (...) We embrace innovation — new ingredients, techniques, appliances, information, and ideas — whenever it can make a real contribution to our cooking. We do not pursue novelty for its own sake. We may use modern thickeners, sugar substitutes, enzymes, liquid nitrogen, *sous-vide*, dehydration, and other non-traditional means, but these do not define our cooking. These are a few of the many tools that we are fortunate to have available as we strive to make delicious and stimulating dishes. Similarly, the disciplines of food chemistry and food technology are valuable sources of information and ideas for all cooks. The fashionable term "molecular

gastronomy" (...) does not describe our cooking, or indeed any
style of cooking. (Italics mine)[46]

This foundation seems to be shared at least in part by Davide Cassi,
physicist and director of the culinary physics laboratory of the University of Parma, and co-author with chef Bocchia of the *Manifesto della
Cucina Molecolare Italiana (Manifesto of Italian Molecular Cooking)*
who claims that the starting point is the need for "every innovation to
expand, not destroy the Italian culinary tradition" (Cassi and Bocchia,
2005; cf. also Cassi, 2001). For Cassi, "the real revolution does not con-
sist in observing cooking from a scientific point of view, but in seeing
science from a culinary point of view. To think that the beautiful and
the tasty are describable scientifically is one of the typical follies of scientism". The question is not to "verify if a culinary procedure is correct
scientifically, but rather to establish whether a scientific idea makes any
culinary sense. It has to then integrate and harmonise with all the other
possible points of view: the aesthetic, the nutritional and the cultural
ones, as well as the ethical one, although they all have to be subordinated
to the culinary point of view which we are dealing with".[47]

The meaning of the encounter between science and cooking is
therefore completely overturned: cooking now has its very own development trajectory, an evolution of culinary knowledge that is actually
"cumulative and the result of collective inventions" in a perfect analogy
with the organisational and institutional identity of modern science
incarnated by the aphorism that is often attributed to Newton, "If I
have seen further it is by standing on the shoulders of Giants".[48] Now
cooking, independently and coherently with its own objectives — that
are different and not subordinated to those of science — takes advantage
of the instruments made available by science and technology. There is
therefore no automatism that imposes that food preparation be adapted

---

[46] https://www.theguardian.com/uk/2006/dec/10/foodanddrink.obsfoodmonthly.

[47] Interview by Andrea Grignaffini, http://www.identitagolose.it/sito/it/protagonisti.
php?id_cat=6&id_art=463. Cf. also http://blog.moebiusonline.eu/index.php/2009/10/06/
dialogos-de-cocina/.

[48] Merton (1965) traces the aphorism back to Bernard of Chartres and the 12th
century.

to the developments of science, nor is there any recourse to the authority of science as a source for legitimisation. Cooking asserts its cultural autonomy and rejects any interference from science, even in its terminology, claiming its right to bend science to its own objectives, to make use of it opportunistically, and to "cherry pick" its most useful results without having to recognise its role and even less its cultural supremacy in return.

This attitude reflects more general contemporary dynamics of the relationship between society and science. Public perception of science is increasingly ambivalent, compared to the times of Artusi. Today science is a seductive reference for cookery, yet it can also burn: if one gets too close, one risks being associated with certain perceived negative implications of science, or being seen as opposing "nature" and "tradition". One example is the debate on the use of additives by some chefs, fueled by a series of television programmes between 2009 and 2010 and by the decision of the Italian Health Ministry to ban certain substances.[49]

The same remarkable, surprising reaction by the public when welcoming the results and applications of science for example in the biomedical field, does not seem to be the result of an internalisation of scientific contents and methods, but a more pragmatic — and even opportunistic — reaction. High expectations from a practical point of view do not correspond to an effective inclination to consider science a true cultural reference.[50]

The distance of Adrià's and his colleagues' "new cooking" from the fervent scientism that marks the triumphant entry of science into the kitchen between the XIXth and XXth century, is not too different from the relationship with science that can be found in literature today. Among contemporary authors, there is a frequent use of themes and inspirations linked to science and technology: from novels that have

---

[49] Italian Molecular Cooker "ban" condemned, *Nature*, 16th March 2010, http://www.nature.com/news/2010/100316/full/news.2010.125.html. On this topic also cf. documents by Dario Bressanini on his blog, http://bressanini-lescienze.blogautore.espresso.repubblica.it/.

[50] I have discussed these themes more in depth in Bucchi (2009 and 2010a). For an analysis based on data on the public perception of science, cf. Beltrame and Bucchi (2010).

scientists as their main characters whose human experiences and narratives go way beyond any scientific content, as in the case of *Solar* by Ian McEwan, to works in which the plot is dictated by a theme or a scientific result that is real or potential, the consequences of which trigger a series of events and scenarios, like in the case of *Never Let Me Go* by Kazuo Ishiguro.

On the one hand, today's novels seem to be fascinated not so much by scientific themes strictly speaking, but rather by their consequences and their social and ethical implications, by the broader debates that permeate public discussion — cloning, genetics, climate change. On the other hand, a more cynical point of view might consider science as a sort of "refuge go-to" for the contemporary writer because it is a source of solid and original content, that can revive plots that are otherwise worn out.

Much less present nowadays is that image of science as a propelling force, a driver of change, a new frontier for man and civilisation that particularly characterises literature between the XIXth and XXth century, and no longer just the literature explicitly imbued with scientific elements. An impulse that led Balzac to define the biologist Cuvier — whose competences greatly influenced his *Peau de Chagrin* — "the greatest poet of our century" and Zola to proclaim almost thirty years before the re-discovery of the Mendel laws, that "heredity has its laws, just like gravitation" (in *La Fortune des Rougon*, 1871). A vision and narrative of the future that turned out to be very attractive and successful, for the first time not grounded in politics or religion, capable of offering to writers such as Jules Verne a new source of inspiration, similar to "what Troy had represented for Homer".[51]

Even when it draws extensively in terms of content and characters, contemporary literature has increasingly — and perhaps irreparably — moved away from that epic, just as the "new cookery" has taken a distanced itself from the impulse to embrace, not just selectively some of the results, but the cultural and institutional model of science.

---

[51] Savinio (1984, p. 122).

Digestif:
Ways of interpreting the relationship between science and cooking

> The point is not who was the province
> and who the empire:
> The point was the fire.
>
> LUCIO BATTISTI and PASQUALE PANELLA, *Hegel*

In the course of the menu offered in these pages, different forms of intersection between science and cooking have emerged more or less explicitly; in my view they are relevant for understanding the wider transformations of the relationships between science and society.

The first form goes back to the origins of modern science. In that context, science is still very close to the practices of daily life. It is in a certain sense "natural" for science to deal with food and its preparation. Being engaged, whether practically or metaphorically, with cooking and other familiar activities, contributes to encouraging the social recognition of science, its understanding and incorporation by the erudite classes. Laboratory boundaries and the contexts of scientific discussions are still uncertain and permeable, and they often extend to domestic spaces as well as convivial ones like coffee rooms.

It is also by distancing itself from similar contexts and more generally by emancipating itself from common sense and daily experience that science gains status, particularly in terms of public perception; that same indifference and distancing from the most basic material necessities including food, especially in some highly significant historical moments, helps to shape the public image of scientists.

As soon as science becomes established from a social point of view as an institution bearing its own relevance and authority, it becomes itself a model by which cooking can be inspired, as is well documented by Brillat-Savarin. Even more strongly, between the XIXth and XXth century, science and its protagonists become on one hand a powerful source of cultural legitimatisation for practices like cooking; on the other hand, cooking is seen as a significant vehicle for a triumphant and accomplished entry of science into society and culture, as well as a context of "domestication" for new scientific aspirations like those expressed during those same years by the female public.

Futurist cooking is quite peculiar in this sense. Here science is intended as a *proxy* for the drive towards modernity and more generally towards novelty. Science must enter the kitchen to support, through its methods and results, the shift in pace that the impending times impose on society; cooking must change like painting and literature. Rhetorically, science is thus opposed to a static and retrograde tradition that must be overcome, in order to replace inept housewives and ignorant cooks with chemistry and for chemistry and laboratories to replace the detested pasta with powder and pills. Finally, molecular gastronomy incarnates a series of movements that appear to be characteristic of the wider contemporary dynamics between science and society.

On one hand, the movement of science towards cooking seems to continue along the path from past centuries: the "scientification" of cooking practices as a natural extension of the increasingly relevant role of science in society from which cooking cannot remain immune. On the other hand, cooking is seen as a seductive opportunity for popularisation, by introducing the public to the contents and objectives of science in a more familiar context and one that is closer to daily life; a sort of prosecution with other means of the "missionary" and paternalistic programme that characterising a large part of the public presentation of science since mid-XXth century.[52]

---

[52] For a more detailed description and a critical analysis of this program, the roots of which go back to some experiences from the previous century, cf., for example Bucchi (2009).

The context of pots and pans, however, remains one that science must handle cautiously, making sure it uses the correct dose of closeness and distance to avoid being assimilated into overly "frivolous" practices unworthy of scientific attention. In this sense, it seems the journey that Norbert Elias considers characteristic of the process of civilisation also applies to science, whereby one should hide from view and "remove behind the scenes of social life" — in the kitchen for example — certain operations that have become "distasteful" such as some practices linked to food preparation.[53] Thus removing oneself increasingly from that "ghastly kitchen" through which Claude Bernard thought it necessary to pass in order to reach the glittering heights of science (cf. chap. *Starter*).

Hence the need, for an ever "civilised" science, to create "technical" terms such as "molecular gastronomy", to avoid that science be "crudely" and more directly linked with the term "cooking" — looked upon suspiciously by the Erice centre. Hence the language register, always readily available, with which the assimilation of science into cooking serves derogatory and even degrading purposes for questionable scientific results and despised scientific adversaries.

Finally, for some theorists of molecular gastronomy, cooking is for science a source of media legitimisation and visibility: if celebrity chefs such as Adrià or Blumenthal use science so liberally, there must be a reason! In this way, the circle of celebrity — or if you prefer Merton's expression, the Matthew effect — is closed: the public visibility of science and its relevance in the kitchen rely on Adrià just as *Science in the Kitchen* by Artusi had relied on the notoriety of Mantegazza.

Cooking does much the same: it draws from the results of science, incorporates its terminology, uses its "ingredients" to renew its inspiration, whilst claiming its own independence and keeping its distances from the potentially more controversial aspects, or those that risk challenging its established social and cultural identity.

---

[53] Elias (1969; Eng. trans. 1978, p. 121).

Thus, by alternating closeness and distance, by accusing each other of reciprocal invasions and projecting upon one another their own responsibilities, science and society engage in the kitchen in that game of thrones that characterises their intersection in our times.

References

Abrahams, M.T.
1998   The Best of Annals of Improbable Research, W.H. Freeman.

Accum, F.
1821   Culinary Chemistry, exhibiting the scientific principles of Cookery, with concise instructions for preparing good and wholesome Pickles, Vinegar, Conserves, Fruit Jellies with observations on the chemical constitution and nutritive qualities of different kinds of food, R. Ackermann, London.

Allen, W.
1972   Yes, But Can the Steam Engine Do This?, in Getting Even, Warner.

Artusi, P.
1891   La Scienza in cucina e l'Arte di mangiar bene. Manuale pratico per le famiglie, Landi, Firenze (ed. by P. Camporesi, Einaudi, Torino, 2001); Eng. trans., Toronto, University of Toronto Press, 2003.

Atkinson, E.
1892   The Science of Nutrition, Damrell & Upham, Boston.

Aubrey, J.
1949   Brief Lives and Other Selected Writings, Scribner, New York.

Babbage, C.
1830   Reflections on the Decline of Science in England, and on Some of Its Causes, B. Fellowes, London.

Babington Macaulay, T.

1837   Lord Bacon, in Critical and Historical Essays contributed to the Edinburgh Review, Longman, London, 1849, p. 372.

Baccus, J.

2004   The Strange Death of Sir Francis Bacon: The Do's and Don'ts of Appellate Advocacy in the WTO, in "Legal Issues of Economic Integration", 31, 1, pp. 13–24.

Bachelard, G.

1938   La formation de l'esprit scientifique. Contribution à une psychanalyse de la connaissance, Vrin, Paris.

Bacon, F.

1623   De Augmentis Scientiarum, in Works, vol. I, ed. by J. Spedding, R.L. Ellis, D.D. Heath, London, 1857.

Barham, P.

2001   The Science of Cooking, Springer, Berlin.

Beccaria, G.L.

2009   Misticanze: parole del gusto, linguaggi del cibo, Garzanti, Milano.

Becker, H.

2011   Review of J.H. Searle, Making the Social World, and P. Bogossian, Fear of Knowledge, in "Science, Technology and Human Values", 36, pp. 273–279.

Belloni, L.

1989   La vera storia della fusione fredda, Rizzoli, Milano.

Bellows, A.J.

1867   The Philosophy of Eating, Hurd & Houghton, New York.

Bercé, Y.M.

1984   Le chaudron et la lancette, Presses de la Renaissance, Paris.

Bernal, J.D.
1953   Science and Industry in the XIXth Century, Routledge, London.

Bernard, B.
1865   Introduction à l'étude de la médecine expérimentale, J.B. Baillière et Fils, Paris; Eng. trans. An Introduction to the Study of Experimental Medicine, Dover Publications, New York, 1957.

Bernardi, W.
2005   La cioccolata del Granduca. Il dibattito sul 'nettare messicano' nella Toscana del Seicento, in P. Scapecchi e L. Nencetti (eds.), Cioccolata, squisita gentilezza, Vallecchi, Firenze, pp. 17–44.

Blot, P.
1867   Handbook of Practical Cookery for Ladies and Professional Cooks, D. Appleton and Company, New York.

Blunt, W.
1971   The Complete Naturalist: A Life of Linnaeus, Collins, London.

Bondi, H.
1962   Relativity and Common Sense: A New Approach to Einstein, Anchor Books, Garden City — New York.

Boni, F.
2002   Il corpo mediale del leader, Meltemi, Roma.

Bresadola M. and Cardinali, S.
2009   Dalla tazzina del diavolo al mondo in una tazza, in "Castelli di Yale", X, pp. 115–131, http://cyonline.unife.it/article/view/2057

Brillat-Savarin, A.
1825   Physiologie du gout ou, Méditations de gastronomie transcendante, Charpentier, Paris; Eng. trans. Lindsay & Blakiston, Philadelphia, 1854.

Broadbent, A.

1900   Science in the Daily Meal, T. Burleigh, London.

Browne, J.

1998   I could have retched all night: Charles Darwin and his body, in C. Lawrence and S. Shapin (ed.), Science Incarnate: Historical Embodiments of Natural Knowledge, University of Chicago Press, Chicago, pp. 240–87.

Bucchi, M.

1996   La scienza e i mass media: la fusione fredda nei quotidiani italiani, in "Nuncius", 2, pp. 581–611.

1997   The Public Science of Louis Pasteur: The Experiment on Anthrax Vaccine in the Popular Press of the Time, in "History and Philosophy of the Life Sciences", 19, pp. 181–209.

1998a   Science and the Media. Alternative Routes in Scientific Communication, Routledge, London and New York.

1998b   La scienza imbavagliata. Eresia e censura nel caso AIDS. Limina, Arezzo.

2004   Science in Society. An Introduction to Social Studies of Science, Routledge, London and New York.

2009   Beyond Technocracy. Citizens, Politics, Technoscience, Springer, New York.

2010   Scientisti e antiscientisti. Perché scienza e società non si capiscono, Il Mulino, Bologna.

2011   Introduzione a R.K. Merton, Scienza, religione e politica, Il Mulino, Bologna, 2011.

Bucchi, M. and Lorenzet, A.

2008   Il lato controverso della scienza, in "Nova review- il Sole 24 ore", 3, pp. 29–40.

Cadeddu, A.

1985   Pasteur et le choléra des poules: révision critique d'un récit historique, in "History and Philosophy of the Life Sciences", 7, pp. 87–104.

Calise, M.

2000   Il partito personale, Laterza, Roma-Bari.

Callon, M.

1981   Pour une sociologie des controverses tecnologiques, in "Fundamenta scientiae", 2, pp. 381–99.

Cameron, I. and Edge, D.

1979   Scientific Images and Their Social Uses, Butterworths, London.

Camporesi, P.

1990   Il brodo indiano: edonismo ed esotismo nel Settecento, Garzanti, Milano, nuova ed. 1998.

1994   The Anatomy of the Senses. Natural Symbols in Medieval and Early Modern Italy, Polity Press, Cambridge.

Capatti, A. and Montanari M.

1999   La cucina italiana, storia di una cultura, Laterza, Roma-Bari.

Cassi, D.

2011   Science and cooking: the era of molecular cuisine, in "EMBO reports", 12, pp. 191–196.

Cavazza, M.

1979   Metafore venatorie e paradigmi indiziarii nella fondazione della scienza sperimentale, in "Annali dell'Istituto di discipline filosofiche dell'Università di Bologna", 1 (1979–1980), pp. 107–133.

Ceresa, M.

1993   La scoperta dell'acqua calda, Leonardo, Milano.

Chamfort (Sebastien-Roch Nicolas)

1795   Produits de la Civilisation perfectionnée, Eng. trans. Selected Essays (Easton Studio Press, 2014).

Chapman, A.

2005   England's Leonardo, Robert Hooke and 17th Century's Revolution, IOP, Bristol.

Cock, T.

1675   Kitchin-physick, or, Advice to the poor by way of dialogue betwixt Philanthropos, physician, Eugenius, apthecary [sic], Lazarus, patient: with rules and directions, how to prevent sickness, and cure diseases by diet, Dorman Newman, London.

Collins, H.M. and Pinch, T.

1974   The TEA Set: Tacit Knowledge and Scientific Networks, in "Science Studies", 4, pp. 165–186.

1993   The Golem: What Everyone Should Know about Science, Cambridge University Press, Cambridge, UK and New York; trad. it. Il Golem. Tutto quello che dovremmo sapere sulla scienza, Dedalo, Bari, 1995.

Collins, H.M. and Yearley, S.

1992   Epistemological Chicken, in A. Pickering, Science as Practice and Culture, Chicago University Press, Chicago and London.

Conley, E.

1914   Principles of Cooking: A Textbook in Domestic Science, American Book Co., New York.

Darmon, P.

1986   La longue traque de la variole, Perrin, Paris.

De Kruif, P.

1926   Microbe Hunters, Harcourt Brace and Company, .

Delbourgo, J.

2011   Sir Hans Sloane's Milk Chocolate and the Whole History of the Cacao, in "Social Text", 29, 1, 106, pp. 71–101.

Di Stefano, A.

2004   Agnello, cacio marzolino e crema battuta: dal menù delle favole settecentesche, in La sapida eloquenza. Retorica del cibo e cibo retorico, ed. by Cristiano Spila (Atti del convegno, Roma 27-28 maggio 2003), Roma, Bulzoni, 2004, pp. 175–191.

Dubos, R.

1960   Pasteur and Modern Science, Doubleday, New York.

Dubourcq, H.

2000   Benjamin Franklin Book of Recipes, Bath, Campus.

Eamon W.

1994a   Science and the Secrets of Nature: Books of Secrets in Medieval and Early Modern Culture, Princeton University Press, Princeton.

1994b   Science as a Hunt, in "Physis", XXXi, 2, pp. 393–432.

Elias, N.

1969   Über den Prozeß der Zivilisation. Band 1: Wandlungen des Verhaltens in den weltlichen Oberschichten des Abendlandes, Suhrkamp, Frankfurt; trad. it. La civiltà delle buone maniere, Il Mulino, Bologna, 1982.

Felici, G.B.

1728   Parere intorno all'uso della cioccolata scritto in una lettera dal conte dottor Gio.Batista Felici All'illustriss. Signora Lisabatta Girolami d'Ambra, Giuseppe Manni, Firenze.

Finlay, M.R.

1992   Quackery and Cookery: Justus von Liebig's Extract of Meat and the Theory of Nutrition in the Victorian Age, in "Bulletin of the History of Medicine", 66, 3, pp. 404–418.

1995   Early Marketing of the Theory of Nutrition: The Science and Culture of Liebig's Extract of Meat, in H. Kamminga and A. Cunningham (eds.), The Science and Culture of nutrition, 1840–1940, Rodopi, Amsterdam, pp. 48–74.

Flaubert

1881   Bouvard et Pécuchet, Lemerre, Paris; Eng. trans. Complete Works of Gustave Flaubert, Delphi Classics, 2013.

Franklin, B.

1868   The Autobiography of Benjamin Franklin, ed. by J. Bigelow, Lippincott, Philadelphia.

Franks, F.

1981   Polywater, MIT Press, Cambridge (Mass.).

Freud, S.

1905   Der Witz und Seine Beziehung zum Umbewussten, Deuticke, Leipzig-Wien; Eng. trans. Wit and Its Relation to the Unconscious, Fisher Unwin, London, 1917.

Garfinkel, H.

1956   Conditions of successful degradation ceremonies, in "American Journal of Sociology", 61, pp. 421–422.

Garzoni, T.

1585   La Piazza universale di tutte le professioni del mondo, appresso Gio. Battista Somalco, Venezia, ed. 1588.

Geison, G.L.

1995   The Private Science of Louis Pasteur, Princeton University Press, Princeton.

Giacchi, O.
1882   Il medico in cucina, ovvero perché si mangia e come dobbiamo mangiare, Emilio Croci, Milano.

Gieryn, T.F. and Figert, A.E.
1990   Ingredients for a Theory of Science in Society, in S.E. Cozzens
and T.F. Gieryn (eds.), Theories of Science and Society, Indiana
University Press, Indianapolis, IN, pp. 67–97.

Giusti, E.
2004   La matematica in cucina, Bollati Boringhieri, Torino.

Goffman, E.
1961   Encounters: Two Studies in the Sociology of Interaction,
Bobbs-Merill, Indianapolis.

Goodwin, C.
2003   Il senso del vedere, Meltemi, Roma.

Govoni, P.
2002   Un pubblico per la scienza: la divulgazione scientifica
nell'Italia in formazione, Carocci, Roma.

Gribbin, J.
2007   The Fellowship: Gilbert, Bacon, Harvey, Wren, Newton, and
the Story of a Scientific evolution, Overlook, London.

Guerrini, A.
2016   The Ghastly Kitchen, in "History of Science" Vol. 54(1),
71–97.

Harford, T.
2006   The Undercover Economist, new ed., 2012.

Imperiale, G.
1663   Le notti beriche, ouero de' quesiti e discorsi fisici, medici,
politici, historici e sacri, Paolo Baglioni, Venezia.

Irace, E.

2003   Itale Glorie, Il Mulino, Bologna.

Jacob, M.C. and Stewart, L.

2004   Practical Matter, Newton's Science in the Service of Industry and Empire, 1687–1851, Harvard University Press, Cambridge, MA and London.

Jardine, L. and Stewart, A.

1998   Hostage to Fortune: The Troubled Life of Francis Bacon, Hill and Wang, New York.

Kantorowicz, E.H.

1957   The King's Two Bodies. Study in Medieval Political Theology, Princeton University Press, Princeton.

Kellogg, E.E.

1892   Science in the Kitchen. A scientific treatise on food substances and their dietetic properties together with a practical explanation of the principles of healthful cookery, Health Publishing Company, Battle Creek.

Klencke, H.

1857   Chemisches Koch- und WirtschaftsBuch, Eduard Rummer, Leipzig.

Inwood, S.

2004   Coffee Shop Society in 17th Century London, Lecture at Gresham College, 25 October 2004.

Lanza, R. and Klimanskaya, I.

2009   (eds.) Essential Stem Cell Methods, Academic Press, Oxford.

Lapucci, C.

1979   Dizionario dei modi di dire della lingua italiana, Garzanti, Milano.

Latour, B.

1992   The costly ghastly kitchen, in A. Cunningham and P. Williams (eds.), The Laboratory Revolution in Medicine, Cambridge University Press, Cambridge, pp. 295–303.

1997   Socrates' and Callicles' Settlement — or, the Invention of the Impossible Body Politic, in "Configurations", 5, pp. 189–240.

Lewenstein, B.V.

1992   Cold fusion and hot history, in "Osiris", second series, 7, pp. 135–163.

Lorenzet, A.

2011   Controversie scientifico-tecnologiche, in M. Bucchi and A. Bonaccorsi, Trasformare conoscenza, trasferire tecnologia. Dizionario critico delle scienze sociali sulla valorizzazione della conoscenza, Marsilio, pp. 86–88.

Malinowski, B.

1948   Magic, Science and Religion, Free Press, New York, 1976.

Mangione, D.

2003   Retorica, critica e ragione del gusto in "La sapida eloquenza. Retorica del cibo e cibo retorico", ed. by C. Spila, Bulzoni, Roma, 2004, pp. 21–34.

Mantegazza, P.

1854   Fisiologia del piacere, Tip. G. Bernardoni, Milano.

1866   Enciclopedia igienica popolare. Igiene della cucina, Brigola, Milano.

Marinetti, F.T. and Fillìa

1932   La cucina futurista, Sonzogno, Milano, Eng. trans. The futurist cookbook, Bedford Arts, San Francisco, 1989.

Markham, G.

1615   The English Hus-wife, Contayning, The inward and outward vertues which ought to be in a compleat woman. As, her skill in Physicke, Cookery, Banquetingstuffe, Distillation, Perfumes, Wooll, Hemp, Flax, Dayries, Brewing, Baking, and all other things belonging to an Houshould, John Beale, London.

Martin, B. and Richards, E.

1995 Scientific knowledge, controversy and public decision-making, in Jasanoff, S. et al. (eds.), Handbook of Science and Technology Studies, Sage, Thousand Oaks, pp. 506–526.

McGee, H.

1984 On Food and Cooking, The Science and Lore of the Kitchen, Scribner, New York.

Méró, L.

1998 Mindenki maskepp egyforma, Tericum Kiadó, Budapest, Eng. trans. Moral Calculations: Game Theory, Logic, and Human Frailty (Springer-Verlag, NY, 1998).

Merton, R.K.

1965 On the Shoulders of Giants. A Shandean Postcript, Free Press, New York; 2nd ed. Orlando, Harcourt Brace, 1985.

1973 The Sociology of Science, The University of Chicago Press, Chicago.

2011 Scienza, religione e politica, ed. by M. Bucchi, Il Mulino, Bologna.

Merton, R.K. and Barber, E.G.

2004 The Travels and Adventures of Serendipity, A Study in Sociological Semantics and the Sociology of Science, Princeton University Press, Princeton.

Montanari, M.

2009 Il riposo della polpetta e altre storie intorno al cibo, Laterza, Roma.

Moseley, B.

1785   Observations on the Properties and Effects of Coffee, John Stockdale, London.

Mulkay, M.

1974   Conceptual Displacement and Migration in Science: A prefatory paper, in "Science Studies", IV, pp. 205–234.

1988   On Humour: Its Nature and Its Place in Modern Society, Polity Press.

Mulkay, M. and Gilbert, N.

1982   Joking Apart: Some Recommendations Concerning the Social Study of Science, in "Social Studies of Science", 12, pp. 585–614.

Pasteur, L.

1939   Oeuvres complètes, réunies par Louis Pasteur Vallery-Radot, VII voll., Masson, Paris.

Perec, G.

1996   Experimental Demonstration of the Tomatotopic Organization in the Soprano (*Cantatrix Sopranica L.*), trad. Eng. Sciences Po University Press, Paris, 1996.

Pettini, A.

1905   Dall'empirismo alla cucina scientifica, Bracciano, Tipografia Romana.

Pignotti, L.

1823   Poesie di Lorenzo Pignotti aretino, Marchini, Firenze (1st ed., Favole e novelle, Pisa, 1782).

Pinch, T.

1992   Opening Black Boxes: Sciences, Technology and Society, in "Social Studies of Science", 22, 3, pp. 487–510.

1997   What's Cooking, in "Mercury", January/February: 29–31.

Pinch, T. and Collins, H.

1984   Private Science and Public Knowledge: The Committee for the Scientific Investigation of the Claims of the Paranormal and Its Use of the Literature, in "Social Studies of Science", 14, 521–46.

Plato

1967   Gorgias, in Plato in Twelve Volumes: With an English Translation, Harvard University Press.

Raichvarg, D. and Jacques, J.

1991   Savants et ignorants: une histoire de la vulgarisation des sciences, Seuil, Paris.

de Réaumur, R.A.F.

1749   Art de faire eclorre en toute saison des oiseaux domestiques, 3 volumi, Paris. Trad. Eng. The Art of Hatching and Bringing up Domestic Fowls of all Kinds at any Time of the Year (C. Davis, 1750).

Redi F.

1809–1811   Opere, Tipografia de' Classici Italiani, Milano, IX Vol.

Ricci, G.

1998   Il principe e la morte. Corpo, cuore, effigie nel Rinascimento, Il Mulino, Bologna.

Richards, E.H.

1882   The Chemistry of Cooking and Cleaning, Estes & Lauriat, Boston.

Rima D. Apple

1995   Science gendered: Nutrition in the United States, 1840–1940, in H. Kamminga and A. Cunningham (ed.), The Science and Culture of Nutrition, 1840–1940, Rodopi, Amsterdam, pp. 129–154.

Rossi, P.

1962   I filosofi e le macchine, Feltrinelli, Milano (nuova edizione 2002).

1991   Introduzione a Bacone. La nuova Atlantide, TEA, Milano.

Rudwick, M.

1975   Caricature as a Source for the History of Science: De la Beche's Anti-Lyellan sketches of 1831, in "Isis", 66, 234, pp. 534–560.

Russell, B.

1957   The Problems of Philosophy; new ed. Oxford University Press, 1998.

Sagredo (Rinaldo De Benedetti)

1960   Aneddotica delle scienze, 2^ed. ampliata, Hoepli, Milano.

Savinio, A.

1984   Narrate, uomini, la vostra storia, Adelphi, Milano.

Schaffer, S.

2005   Public Experiments, in B. Latour (ed.), Making Things Public, ZKM/Mit Press, Cambridge, MA, pp. 298–307.

Schivelbusch, W.

1980   Das Paradies, der Geschmack und die Vernunft. Eine Geschichte der Genußmittel, Hanser, Munchen; trad. Eng.

Schlipp, P.A. (ed.)

1949   Albert Einstein. Philosopher-Scientist, vol. III, Evanstone.

Seeling, T.L.

1991   The Epicurean Laboratory: Exploring the science of cooking, Freeman/Scientific American, New York.

1994   Incredible Edible Science, Freeman/Scientific American, New York.

Shapin, S.

1998   The Philosopher and the Chicken: On the Dietetics of Disembodied Knowledge, in S. Shapin and C. Lawrence (eds.), Science Incarnate. Historical Embodiments of Natural Knowledge, University of Chicago Press, Chicago, pp. 21–50.

2000   Descartes the Doctor: Rationalism and Its Therapies, in "The British Journal for the History of Science", 33, pp. 131–154.

2008   The Scientific Life. A Moral History of a Late Modern Vocation, The University of Chicago, Chicago and London.

Shapin, S. and Schaffer, S.

1985   Leviathan and the Air-pump. Hobbes, Boyle, and the Experimental Life, Princeton University Press, Princeton.

Sloane, H.

1707–1725   Voyage to the Islands Madera, Barbados, Nieves, St Christophers and Jamaica, with the Natural History of the Herbs and Trees, Four-footed Beasts, Fishes, Birds, Insects, Reptiles, etc. of the last of those Islands, of Natural History of Jamaica, printed for the author, London.

Söderlind, U.

2005   Nobels middagar: banketter, festligheter och pristagare under 100 år, Carlsson, Stockholm; trad. Eng. The Nobel Banquets: A Century of Culinary History (1901–2001), World Scientific, 2010.

Snodgrass, M.E.

2004   Encyclopedia of Kitchen History, Taylor & Francis, London and New York.

Standage, T.

2005   A History of the World in Six Glasses, Walker & Company, New York.

Strand Noad, S.

1979   Recipes for Science Fun, Watts, New York.

Stubbe, H.

1662   The Indian Nectar, or a Discourse Concerning Chocolata, J.C. Fdor Andrew Crook, London.

Sturgeon, L.

1822   Essays, Moral, Philosophical, and Stomachical, on the Important Science of Good Living, Whittaker, London.

Taleb, N.N.

2007   The Black Swan, Random House, New York.

Thelle, D.S. and Strandhagen, E.

2005   Coffee and disease: an overview with main emphasis on blood lipids and homocysteine, in "Scandinavian Journal of Nutrition", 49, 2, pp. 50–61.

This, H.

1993   Les secrets de la casserole, Belin, Paris; Eng. trans. Kitchen Mysteries, Columbia University Press, New York.

2006   Food for tomorrow? How the scientific discipline of molecular gastronomy could change the way we eat, in "EMBO reports", 7, pp. 1062–1066.

Tissot, S.A.A.D.

1768   De la santé des gens des lettres, Losanna, François Graiset, Eng. trans. An Essay on Diseases Incidental to Literary and Sedentary Persons: With Proper Rule for Preventing Their Fatal Consequences and Instructions for Their Cure, Edward and Charles Dilly, London, 1768.

Turbil, C.

2017   Paolo Mantegazza and the dream of 'making' science popular circa 1860–1900, in "Public Understanding of Science", vol. 26, 5, pp. 627–631.

Vallery-Radot, R.

1900   La vie de Pasteur, Flammarion, Paris (30eme edition, 1931, Hachette, Paris); René Vallery-Radot, The Life of Pasteur, Vol. II (Constable & Co., London, 1911).

Vallortigara, G.

2005   Cervello di Gallina. Visite (guidate) tra etologia e neuro-scienze, Bollati Boringhieri, Torino.

Verri, P.

1766   Il Caffè: o sia, Brevi e varj discorsi già distribuiti in fogli, Venezia.

Waddington, K.

2010   More like Cooking than Science: Narrating the Inside of the British Medical Laboratory, 1880–1914, in "Journal of Literature and Science", 3, 1, pp. 50–70.

Walsh, J.H.

1856   A Manual of Domestic Economy, Routledge & Co., London.

Watson, J.

1968   The Double Helix: A Personal Account of the Discovery of the Structure of DNA, Atheneum, New York.

Waxter, J.B.

1981   The Science Cookbook: Experiment recipes that teach science and nutrition, Fearon, Belmont.

Werrett, S.

2000   Healing the Nation's Wounds: Royal Ritual and Experimental Philosophy in Restoration England, in "History of Science", 38, pp. 377–399.

Williams, W.M.

1885   The Chemistry of Cookery, Appleton, New York.

Willis, T.

1685   The London Practice of physick, or The whole practical part of physick, London.

Wolke, R.L.

2002   What Einstein Told His Cook, Norton, New York.

Woodward, K. and Heddle, R.

1992   Science in the Kitchen, EDC, London.

Zeti, F.

1728   Altro parere intorno alla natura, e all'uso della cioccolata, Firenze.

Index of Ingredients

Acknowledgements

To Gabriele Bucchi, Marco Cavalli, Serena Luzzi and Renato G. Mazzolini, for their careful reading and useful suggestions.

To Giuseppe Pellegrini and Barbara Saracino, for their comments on preliminary versions of the manuscript.

To Eliana Fattorini and Susan Howard for their help in revising text and references.

To Emanuela Minnai, for her patient and wise encouragement.

To Davide Cassi and Monica Cioli, for some important information and reference suggestions for the chapter *Dessert*.

To Patricia Osseweijer and all the colleagues at the Technische Universiteit Delft, for hosting me as visiting professor during part of the research work for this book.

To Mario Bagnara, Cecilia Magnabosco and all the staff of the Biblioteca Internazionale La Vigna and of the Wellcome Library.

To all the students and colleagues attending my seminars at the University of Nijmegen, York (Canada) and the Medical Museion, Copenhagen.

Conversations with the late Carlo Cannella have been a source of rich inspiration for this work.

Printed in the United States
by Baker & Taylor Publisher Services

Printed in the United States
by Baker & Taylor Publisher Services